W9-AIB-053

Crossing
the Deadly
Ground

Crossing

356.183
J 242

the Deadly Ground

United States Army Tactics, 1865-1899

Perry D. Jamieson

Howard County Library
Big Spring, Texas 79720

THE UNIVERSITY OF ALABAMA PRESS

TUSCALOOSA & LONDON

Copyright © 1994
The University of Alabama Press
Tuscaloosa, Alabama 35487-0380
All rights reserved
Manufactured in the United States of America
designed by Paula C. Dennis

The paper on which this book is printed meets the minimum
requirements of American Standard for Information Science-
Permanence of Paper for Printed Library Materials,
ANSI Z39.48-1984.

Library of Congress Cataloging-in-Publication data
Jamieson, Perry D.
Crossing the deadly ground : United States Army tactics,
1865–1899 / Perry D. Jamieson.
p.   cm.
Includes bibliographical references and index.
ISBN 0-8173-0720-6 (alk. paper)
1. United States. Army. Infantry—Drill and tactics—History—
19th century. I. Title.
UD160.J36   1994
356'.183—dc20   93-45318

British Library Cataloguing-in-Publication Data available

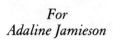

For
Adaline Jamieson

Contents

Illustrations

Preface

Military historians have written countless volumes about the Civil War, World War II, and other major American wars but relatively few books about the United States Army during the late nineteenth century. The great campaigns and bloody battles of our history have attracted far more attention than have the last decades of the 1800s, a period of peace and limited conflicts.

The United States Army did not fight any large conventional wars between 1865 and 1917. It waged a series of skirmishes and small battles against the western Indians, opponents who rarely used tactics resembling those of European armies. The Spanish-American War, the only entirely conventional conflict during this period, ended rapidly and without much bloodshed. Its sequel, the Philippine War, began with traditional operations, but soon deteriorated into guerrilla warfare.

Although the late nineteenth century brought the United States Army no major conventional conflicts, it nonetheless proved a crucial period in American military history. Service associations appeared and professional publications such as the *Army and Navy Journal*, the *Army and Navy Register*, and the *Journal of the Military Service Institution of the United States*, flourished. The army emphasized military education: the Artillery School was revived in 1868, the Infantry and Cavalry School was established in 1881, and the Engineer School emerged about the same time. Lively discussions took place in the service's classrooms and journals, for the late nineteenth century was also an era when new weapons created controversies about organization, training, strategy, and tactics.

The tactical problem that dominated the period began to emerge during the Civil War, when defenders protected by field works delivered rifled infantry fire and deadly artillery blasts against attackers approaching in close-ordered lines. After the ghastly battles of the 1860s, improvements in weapons technology and field engineering made assaults even more dangerous than they had been for General George E. Pickett's men at Gettysburg or John Bell Hood's at Franklin. How could attackers advance across the open terrain in front of defenders who were so well armed and protected? In 1882, while many American officers pondered this dilemma, a British theorist described it this way: "A certain space of from 1,500 to 2,500 yards swept by fire, the intensity of which increases as troops approach the position from which that fire is delivered, has to be passed over. How shall it be crossed?"[1]

During the late nineteenth century, thoughtful American soldiers struggled with this frustrating challenge and others. They argued about the answers to their problems, prepared new tactical manuals, sharpened their marksmanship, conducted field exercises, and suffered in combat. The ideas they debated during peacetime and the tactics they used in battle help us understand a larger story, the journey of the United States Army into the twentieth century.

Acknowledgments

Many people contributed to this book over several years and if I fail to mention all of them here, their help is no less appreciated.

Grady McWhiney of Texas Christian University and Russell F. Weigley of Temple University deserve particular thanks for their advice and encouragement. I am also indebted to an anonymous referee who reviewed my manuscript and to Robert M. Utley and Graham Cosmas, who read parts of it. These scholars offered many useful suggestions.

Jeff Slannery helped my research at the Library of Congress and Sylvan DuBow, William Grace, and others assisted me at the National Archives.

No student can finish any serious project in late nineteenth-century American military history without the aid of the archivists and librarians at the United States Army Military History Institute. I am particularly grateful to Richard J. Sommers, David Keough, Pamela Cheney, John Slonaker, Louise Arnold-Friend, Dennis Vetock, and the other superb professionals who work in Upton Hall.

Several people offered useful research leads, read some chapters of my manuscript, located photographs, ordered books through inter-library loan, or otherwise encouraged me: Barbara Auman, Rick Eiserman, Randy Hackenburg, Yvonne and Joe Kincaid, Judith Schafer, Richard J. Sommers, and Michael Winey.

The University of Alabama Press strongly supported this project, from start to finish. I particularly appreciate the encouragement of Malcolm MacDonald, the copyediting of Marcia Brubeck, and the help of Judith Knight, Anders L. Thompson, Suzette Griffith, and others at the Press.

My wife Stephanie Jamieson kept her good humor throughout the years I worked on this book. I am thankful for her patience during the many evenings and weekends that I gave to the late nineteenth-century army, and not to her.

My friends Paul Charette and Suellen Wood played an important role in this project. Every few months one of them would ask, "When are you going to finish that book of yours?"

Perry D. Jamieson
Crofton, Maryland

Crossing
the Deadly
Ground

I

No More Cold Harbors
Issues in Tactics, 1865–1880

The chaplain of the First Massachusetts discerned a pattern in the confused, back-and-forth fighting that he saw during the Battle of the Wilderness. "Wherever the Federal troops moved forward," he observed, "the Rebels appeared to have the advantage. Whenever they advanced, the advantage was transferred to us." This New England clergyman, who witnessed several attacks during the gruesome combat that opened the Virginia campaign of 1864, had identified the fundamental tactical problem of the Civil War: defending troops usually held the upper hand over attackers. The introduction of rifled shoulder arms, the power of artillery on the defensive, and the use of field entrenchments, which became more sophisticated during the course of the war, all weighed heavily on the side of the defense. Civil War attacks, frontal assaults in particular, often ended in bloody failures. The most famous Confederate leader, General Robert E. Lee, suffered his worst defeats while fighting on the tactical offensive, at Malvern Hill and at Gettysburg. One of the Union's most inept commanders, Major General Ambrose E. Burnside, lost more than 12,000 soldiers dur-

ing a single day of senseless attacks at Fredericksburg, and one of its greatest heroes, Lieutenant General Ulysses S. Grant, sent nearly 14,000 of his
men to slaughter in two days at Cold Harbor.[1]

The strengths of the tactical defensive and the large size of Civil War
armies made them extremely resilient, able to absorb hard blows, withstand
heavy casualties, and still keep their cohesion. An army, or part of it, might
be routed from a battlefield, the fate at First Bull Run of Brigadier
General Irvin McDowell's collection of poorly trained divisions or at
Chancellorsville of Major General Oliver O. Howard's ill-starred Eleventh
Corps. But during the entire four years of the Civil War, only one large field
force can be said to have been destroyed in combat. At the Battle of Nashville on December 15, 1864, well-executed Union attacks drove in both
flanks of the Army of Tennessee. Its survivors fled to Mississippi and never
again formed an effective army.

This lone exception to the resilience of Civil War armies is readily explained by events two weeks earlier at the Battle of Franklin, where General John Bell Hood, the most aggressive of Southern army commanders,
sacrificed the Army of Tennessee in a mindless frontal assault that provided
another tragic example of the power of the tactical defensive. In a single
heroic advance, the Confederates suffered total casualties of 7,000 troops, at
least a third of their attacking infantry, losses that included six generals
killed or mortally wounded and an untold number of experienced company and regimental officers. It was hardly remarkable that Hood's army
met the fate that it did at Nashville, half a month after its calamity at
Franklin. After the end of the Civil War, thoughtful soldiers foresaw that
the tactical defensive would become even stronger, and armies even more
resilient, as breech-loading arms—if not repeaters—became more common and that as firepower increased, the value of the bayonet and saber
would diminish.[2]

These developments suggested that the post–Civil War army needed
new ideas about how to fight its future battles, and some officers concluded
that it should develop tactics of its own, independent of European influence. "Now that the war is ended," the *Army and Navy Journal* editorialized
in November 1865, "some of our officers evidently fancy it to be the proper
time to start a new school of warfare which, no doubt, they would call the

'American School.'" Earlier that autumn one Civil War veteran had declared: "We are a practical people. ... Let us leave show and useless, brain-confusing evolutions to monarchial Europe." After the reformer Emory Upton published in 1867 the first tactical manual authorized after the Civil War, the commanding general of the army, Ulysses S. Grant, emphasized to the secretary of war that this work, unlike its predecessors, was "purely American" and not a translation. When the long-tenured veteran Brigadier General John Pope enumerated some of the United States Army's problems in 1873, he included the "feeble imitation of foreign systems."[3]

An "American school" of tactics probably had no stronger advocate in the post–Civil War army than its commanding general, William T. Sherman. In 1874 Sherman reminded his old comrade Stephen A. Hurlbut that throughout their careers the United States military had been shaped by the "principles and practices" of its counterpart in France. "In 1840 the French nation stood preeminent as a military people," Sherman recalled, "and it was natural and proper that we should in a measure be influenced by their example; but the Institutions of every country should harmonize with the genius & tone of the mass of the people. Our People are not French but American, and our Army should be organized and maintained upon a model of our own, and not copied after that of the French, who differ from us essentially." Sherman's nationalism also was evident when, later in the 1870s, he made some suggestions for the curriculum of the Artillery School at Fort Monroe, Virginia, which, since its revival in 1868, had required students to give presentations on military history to their classmates and staff. The commanding general recommended that every artilleryman should research, write, and read to his class a paper on the strategy and tactics of a campaign from American history. Sherman proposed a list of such topics, beginning with the Brandywine campaign of George Washington, the nation's premier military hero, and ending with the operations then still in progress against the Plains Indians. (He also modestly included two case studies from his own career.) At the end of the 1870s, the content of the Artillery School's ten-week course in military history and geography was as nationalistic as the commanding general's recommendations for its student papers. Although the curriculum gave a nod to Marlborough and Frederick the Great, it concentrated on the American Civil War.[4]

Another senior army officer, Lieutenant General Philip H. Sheridan, doubted that Americans could gain much from studying European tactics, an opinion he formed while observing the Franco-Prussian War. "I find that but little can be learned here to benefit our service," he wrote from France during the Sedan campaign. "We are far ahead in skill and Campaign Organization." On another occasion Sheridan reported that he saw "many things of great interest . . . but not much in this Old World which should be taken as a standard for the New World." The peppery American general said of the Franco-Prussian conflict in his memoirs: "I saw no new military principles developed, whether of strategy or grand tactics."[5]

In addition to nationalist pride like Sheridan's, there was at least one practical reason for the United States Army to develop its own tactics in the late 1860s: the likelihood that, at least for the foreseeable future, it would train and fight on American, and not European, terrain. Foreign war was thought so unlikely that the late nineteenth-century army rarely expended its meager funds on training regiments as full units. The *Army and Navy Journal* wondered in 1865 if the time had arrived for the service to embrace tactics tailored to "the nature of our country." When Emory Upton's first tactical manual appeared two years later, it carried the subtitle, "Adapted to American Topography and Improved Fire-Arms."[6]

An "American school" of tactics was also promoted by the belief that the translations of European manuals by William J. Hardee and Silas Casey used during the Civil War had proven inadequate. "We are beginning to overhaul our Scott, our Hardee, our Casey," the *Army and Navy Journal* observed in November 1865, "and to question whether, after all, the officers who edited these tactics did not follow the original too closely." G. K. Warren, who had been a corps commander in the Army of the Potomac, was quoted after the war as complaining that Casey's manual left "the Army in some situations virtually without any tactics at all." Time and again during the war, frontal assaults in the close-ordered lines of Hardee and Casey, against defenders armed with rifles and protected by entrenchments, had ended in disaster. Long after the Civil War, in 1890, one veteran who remembered these bitter experiences passed on this warning: "We can never again march solid first lines to the attack under heavy fire, as at Fredericksburg or Cold Harbor."[7]

Samuel Beatty's brigade at the Battle of Stone's River, deployed in the close-ordered lines typically used during the Civil War. Fighting at the ranges shown here, defenders armed with rifled muskets inflicted enormous casualties on troops attacking in these tight formations.
(From Johnson and Buell, *Battles and Leaders of the Civil War*, vol. 3, page 622)

No officer of the postwar army was more convinced of the need for new ideas than Emory Upton, who had been critical of the leadership and tactics he had seen during the Civil War[8] and had become one of the conflict's most innovative tacticians. Looking for an alternative to the close-ordered linear tactics of Hardee and Casey, Upton experimented with light-column formations at Rappahannock Station on November 7, 1863, and at Spotsylvania Court House on May 10, 1864.[9] The following April he helped plan, then lead, an assault of dismounted cavalrymen armed with Spencer repeaters that carried an entrenched Confederate position at Selma, Alabama. Before the war ended, this intense officer from New York began developing a tactical system of his own, and in January 1866 he wrote to the assistant adjutant general, asking that his tactics be considered by the secretary of war or by a board of officers.[10]

Two such panels studied Upton's system and recommended that it become the army's authorized tactics. The first was a board of officers chaired by Colonel Henry B. Clitz,[11] and the second, larger and more prestigious, was presided over by the commanding general of the army, Ulysses S. Grant. Acting on the endorsements of these committees, the War Department on August 1, 1867, adopted Upton's work as the official tactics for the United States Army. The author published his effort at his own expense, under the title *A New System of Infantry Tactics, Double and Single Rank*.[12]

Within two years of the publication of this manual, Upton assumed the leading role in the army's effort to develop what nineteenth-century soldiers called an "assimilated" tactics, a system with commands and formations that were compatible among infantry, artillery, and cavalry. A board of officers chaired by a distinguished Union veteran, Major General John M. Schofield, undertook this project in 1869. Upton, who had gained experience with all three arms of the service during the Civil War, proposed to the Schofield Board early in its deliberations that his 1867 infantry manual could serve as the basis for an assimilated system. The committee never formally accepted this offer, but the board's findings, presented to the War Department in 1871, included an infantry tactics similar to Upton's.[13]

The War Department did not endorse the results of the Schofield Board. One contemporary observer alleged that complacent reviewers had filed the panel's report in one of the bureaucracy's "pigeon holes." There may

EMORY UPTON.
Drawing on his Civil War experience, this intense student of tactics prepared
the first official drill manuals used by the army after that bloody conflict.
(Photograph courtesy of United States
Army Military History Institute)

WILLIAM T. SHERMAN.

Commanding general of the army from 1869 until 1883, Sherman served as a
mentor to Upton and also had an interest of his own in weapons and tactics.
(Photograph courtesy of United States Army Military History Institute)

have been sound reasons for the War Department's failure to accept the Schofield Board's efforts. One officer claimed that Philip St. George Cooke, a Union trooper who had written a cavalry manual of his own, raised enough criticisms to discredit the panel's mounted tactics. It was also suggested that the Schofield Board's infantry drill was so similar to Upton's that its adoption would infringe his copyright.[14]

This might well have been the end of the assimilation project, but the army's senior officer intervened. William T. Sherman served as a mentor to Upton, encouraging his work on tactics and other reforms. In the case of the assimilated manuals, Sherman had confidence in the studious New Yorker's knowledge of the subject and perhaps also believed that the best resolution of the copyright issue was the adoption of Upton's own work. The commanding general directed his enterprising subordinate and three other officers to prepare an assimilated tactics, based on Upton's own system.[15]

The War Department authorized the results of their efforts, which were published in 1874 in three volumes, one for each arm of the service. These books were not a combined-arms tactics, advising, for example, a battery commander how best to cooperate with infantry. They were, instead, manuals whose commands and formations were compatible among the three arms, so that an officer could move, for instance, from the artillery service to a cavalry regiment and quickly learn the drill of his new unit. Upton and his comrades did not produce a combined-arms tactics, but they were the first board of officers in the army's history to study the three arms together and design a system applicable to all of them. Eight years after the publication of these assimilated volumes, General Sherman pronounced them "all sufficient for my day and generation."[16]

In both the assimilated tactics and his 1867 work, Upton tried to move troops more efficiently than the Civil War manuals had allowed. Looking for more flexibility than he found in Hardee and Casey, Upton made groups of four men, "fours," the basic units of his infantry system, using them to replace the platoons and sections of earlier drill books. Upton's scheme allowed foot soldiers to march in columns composed of "fours," deploy from column into fighting line by "fours," and use this same primary unit to march by the flank, to wheel, or to perform other movements before or during contact with an enemy. A regiment of ten companies, the basic infantry unit, could align and maneuver by collections of "fours," blocks of eight or twelve men.[17]

Upton's other significant innovation was his introduction of single-rank tactics. He expected breechloaders, and repeating breechloaders in particular, to increase the firepower of an infantry regiment so greatly that in some cases it could be deployed in a single, rather than a two-line, formation. Upton pointed to the wartime successes won by dismounted Union cavalrymen armed with seven-shot Spencer rifles, examples that he believed proved "that one rank of men so armed is nearly, if not quite, equal in offensive or defensive power to two ranks armed with the Springfield musket. If this be admitted, a one-rank tactics becomes necessary for a certain proportion of troops, especially those designed to turn or operate on the enemy's flank." Many years after Upton's single-rank tactics were adopted, one infantry officer reflected that with the advent of breechloaders and Gatling and Hotchkiss guns, footsoldiers deployed in the two-line formations of the Civil War became "simply food for gunpowder. The single rank formation is now and will be the only one used in battle, unless, indeed, the line shall become still more attenuated by the introduction of an open order system."[18]

Upton's innovative 1867 manual was well received in some quarters. The Grant Board gave it a strong endorsement, commending its provision for single-line tactics. The *Army and Navy Journal* devoted two lead articles, in September 1866 and February 1867, to favorable publicity for the work. The year after the manual had been adopted as the army's authorized tactics, one Civil War veteran praised the efficiency of Upton's system of "fours," asking rhetorically: "What is simpler than this?"[19]

Despite his innovations, or perhaps because of them, Upton attracted many critics. William H. Morris, who had published a *Field Tactics for Infantry* of his own in 1864, raised the specter that his rival's new tactics were too novel, and Major General Thomas W. Sherman, a veteran of both the Mexican and Civil Wars, warned that Upton's system contained major, perhaps fatal, flaws. One stubborn soldier declared in 1868: "I remain a staunch adherent of the dethroned Casey."[20]

Whether an officer criticized or praised the new tactics depended in part on his beliefs about the relative merits of firepower and shock. Upton introduced his "fours" and single-line tactics in response to the deadly volleys he had experienced on Civil War battlefields and in the expectation that infantry fire would become even more powerful in the future. An implied corollary was that shock tactics, the use of close-ordered lines or heavy columns to overpower a defender with a sudden bayonet charge, would become increasingly dangerous. In 1874 George B. McClellan, who was an unsuccessful field general but an astute observer of military affairs, warned that these traditional formations could not survive the long-range, rapid, and accurate fire delivered by the latest weapons.[21]

Given the dominance of firepower over shock, some officers concluded that the bayonet had lost most of its value. One soldier complained in 1868: "The trouble with the admirers of the bayonet is that they do not reflect on what the breech-loader can accomplish. . . . There is but one way to oppose breech-loaders, namely, with breech-loaders." Discussing weapons with Philip H. Sheridan in 1878, William T. Sherman suggested that the army should recognize that the bayonet and the noncommissioned officer's sword had lost their utility in combat and exchange them for more practical armaments. Sheridan's chief ordnance officer agreed with Sherman and recommended that the bayonet, along with the saber, "be replaced by a more useful weapon and tool."[22]

In spite of such statements, the sentiment persisted that the infantry should not abandon the bayonet. None of the weapon's proponents were more vigorous than Francis J. Lippitt, a veteran of the Mexican and Civil Wars who published a book on tactical theory just after Appomattox. Suggesting tactics for all three arms of the service, Lippitt's recommendation to foot soldiers was that "the proper mode of attack by infantry on infantry is with the bayonet." He ignored the contrary evidence from Civil War experience and declared: "The bayonet is usually more effective than *grape*, *canister*, or *bullets*." Although few officers were so zealously in favor of the bayonet as Lippitt, the edged weapon did not lack defenders. A lieutenant of the Third Infantry wrote, early in the 1870s: "Nobody . . . would . . . for a moment think of depriving the Infantry arm of half its force by taking away the bayonet." William T. Sherman seemed to retreat from his earlier statement to Sheridan when, in October 1879, the conqueror of Atlanta acknowledged that the army would have to retain its traditional edged weapon in some form.[23] If the commanding general of the army appeared to put himself on both sides of this question within two years, the soldier and lawyer Alfred H. Terry straddled the issue in the course of a single report that he signed in 1872. "I think the day of the bayonet has passed away," Terry offered at one point but later on the same page added: "While, however, this is my belief, I do not think it prudent to definitely abandon the weapon until actual practice in war shall demonstrate that it is no longer of use." The army's reluctance to give up this traditional armament was also evidenced when it issued ten thousand trowel bayonets in 1874. Designed by Lieutenant Edmund Rice, this accessory was both an entrenching tool and a close combat weapon. The trowel bayonet allowed infantry to prepare fieldworks more quickly—an important benefit when defending against fire from breechloaders—and at the same time retain the traditional edged weapon of the foot soldier.[24]

While infantrymen discussed the bayonet, cavalrymen debated the saber. Critics of edged weapons contended that rifled shoulder arms had made mounted charges against infantry too dangerous. They could point out that Civil War saber attacks against foot soldiers had been rare and successful ones rarer still. The prospects darkened further when the infantry gained breechloaders. In 1868 the *Army and Navy Journal* declared the day of the saber had ended, arguing that in nine cases out of ten, infantry massacred horsemen who relied on it. Faced with breechloaders, William T. Sherman predicted, "the bold Sabreur must disappear."[25] The Schofield Board, which studied weapons and accoutrements as well as tactics, recommended that cavalrymen be armed with carbines and revolvers, rather than sabers, and a captain of the Sixth Cavalry commented in 1880 that if the troopers practiced well with these firearms, they would have little need of their edged weapons, for either conventional or Indian warfare.[26]

The defenders of the saber, some of whom seemed to regard any criticism of the weapon as an attack on the entire cavalry service, were as adamant as those of the bayonet. In 1871 Major William R. Parnell of the First Cavalry prepared a saber manual that was based on another one published thirty years earlier. A Union cavalry veteran wrote to contradict the *Army and Navy Journal*: "The days of the sabre are *not* over, and never will be, except for those who have no love for horses and no faith in steel; and for such the days of the sabre never dawned." One volunteer cavalryman disparaged the Schofield Board's findings, citing what he believed was an important lesson of the Franco-Prussian War. "If the recommendations of the St. Louis Board are enforced," this trooper predicted, "our regular cavalry will soon become as useless as the French cavalry proved in the war of 1870, always ready to run away. The sabre is as valuable for its moral effect as for its actual execution." Four months after the army's worst defeat during the Indian wars, another cavalryman claimed that George Armstrong Custer would have "given millions" for a hundred sabers at the Little Big Horn.[27] Wesley Merritt, a highly regarded veteran of the Union cavalry, recommended in 1879 that the cavalry keep its carbines and, "above all," its sabers. One of the army's most conscientious students of weapons and tactics, Colonel John C. Kelton, commented the following year that cavalrymen remained divided over the issue, with "perhaps" a majority in favor of the saber over firearms.[28]

Some horse soldiers concluded that the fault was not the weapon itself but its maintenance: the troopers must keep their sabers well honed. Generals David S. Stanley and William W. Averell believed that sharp blades had raised the morale of their soldiers during the Civil War. "Give our troopers the sabre," advised one cavalryman in the mid-1870s. "Sharpen it and teach them its use." George Custer's biographer Frederick J. Whittaker argued in 1871 that dull weapons were among the mounted arm's worst problems. "Sabres are issued blunt enough to ride on to San Francisco . . . ," he complained from his home in New York. "The men lose confidence in the weapon, and prefer the revolver." Whittaker recommended that cavalrymen carry "razors three feet long."[29]

A cavalryman's opinion as to whether he needed a saber or not depended in part on his answer to the larger question of whether he was more likely to fight mounted or dismounted. During the course of the Civil War, troopers had increasingly left their saddles to fight on foot. James H. Wilson, drawing on his extensive experience during that conflict, advised a board of officers on cavalry tactics in 1868 that they give their main attention to the dismounted drill. The commanding general of the army suggested, ten years later, that horses served chiefly to bring cavalrymen rapidly to the battlefield, where the troopers would fight on foot with their infantry comrades.[30]

Traditionalists continued to champion the idea that cavalry should fight from the saddle. Gilbert E. Overton of the Sixth Cavalry, who won a brevet by leading a charge on a Cheyenne village in 1874, described in 1880 what he considered the ideal mounted unit. "In a charge," Overton wrote, "it would rival the Mamelukes, who did fire from their horses; it would prove a wonder to the Indians of the plains, who invariably dismount to deliver their fire." One soldier grumbled in 1875 that the Civil War had made troopers too enamored with dismounted cavalry fighting, and John Bell Hood's 1880 memoirs harshly criticized this style of combat. The aggressive Southerner contended: "A cavalryman *proper* cannot be trained to fight, one day, mounted, the next, dismounted, and then be expected to charge with the impetuosity of one who has been educated in the belief that it is an easy matter to ride over infantry and artillery, and drive them from the field."[31]

In addition to the mounted-dismounted debate, the saber was a consideration in the controversy over whether the cavalry's standard formation should be single or double rank. The War Department had authorized and published a two-rank tactics in 1841. The work proved popular with Civil War commanders and enjoyed several printings during the conflict. In 1861 the War Department also authorized Brigadier General Philip St. George Cooke's *Cavalry Tactics*, a single-rank system. The two schools competed for decades, until the official cavalry manual of 1891 resolved the issue in favor of the single-rank tactics. It is significant that debaters on each side of this question emphasized how well the formation they favored would contribute to the shock of mounted tactics rather than to the firepower of dismounted fighting.[32]

While cavalrymen debated these issues, artillerymen considered a few tactical questions that were unique to their arm. Some observers judged that rifled shoulder arms had brought a decline in artillery's power, relative to infantry. The autumn after the Civil War one correspondent of the *Army and Navy Journal* offered the opinion that field gunnery had "lost its terrors." Six years later, another writer expanded on the point: "The glory of the field artillery has in a measure departed. Batteries lean more heavily on the infantry for support than of old, and there is no longer the scope there once was for 'the judgment, the dash and enterprise, which, in the days of short-ranged and muzzle-loading small arms, went so far in making up the character of the model battery commander.'" A British military observer concluded in the late 1870s that infantry skirmishers could sweep artillerymen from their pieces, while field or even siege guns could make little impression against earthworks staunchly defended by breechloaders.[33]

Still other soldiers believed that the artillery was gaining in importance. George B. McClellan predicted in 1874 that as field batteries replaced their muzzle-loading pieces with breech-loading ones, they would be able to operate with minimal support from the infantry and that their commanders would thus become more independent. Francis J. Lippitt's _A Treatise on Intrenchments_, published in 1866, assigned artillery a significant role in attacks on fieldworks. A year later the _Army and Navy Journal_ emphasized rifled cannon, which had greater accuracy and range than smoothbore pieces, among the factors that were forcing infantry to adopt looser formations.[34]

One of the sharpest artillery debates began with the appearance of the earliest machine guns, which were initially considered fieldpieces rather than infantry weapons because, like cannon, they were mounted on carriages and worked by crews. (The examination for commissioned officers at the Artillery School in 1878 included exercises with "Field guns, including Gatlings.") The first of these new arms were the French Montigny mitrailleuse and the American Ager[35] and Gatling guns. Patented in 1862, the initial model of Richard J. Gatling's weapon had six barrels that were rotated by a hand crank. A later version, tested in January 1865, fired twenty rounds from four barrels in eight seconds.[36] Only a few Civil War officers, all of them Northerners—David D. Porter, Winfield S. Hancock, Benjamin Butler, and John W. Geary—took any interest in the new, rapid-firing weapons.[37] President Abraham Lincoln gave the Ager gun both its nickname, the "Coffee Mill" gun, and his hearty support.[38] But the interest taken by these senior officers and the president was outweighed by resistance from the Union chief of ordnance, James W. Ripley, and the machine gun saw little service during the Civil War.[39]

After Appomattox inventors continued to improve these rapid-firing weapons, and some soldiers welcomed this new ordnance. George B. McClellan found "good grounds for believing" in 1874 "that for the defence of works, of defiles, or of a position of limited extent[,] the mitrailleuse, or, still better, the Gatling gun, will prove to be a very reliable adjunct." Frederick Whittaker predicted that the new weapons would be "invaluable" on the defensive against close-in assaults. William F. Barry, an accomplished artillerist in both theaters of the Civil War, chaired a board of officers that prepared a new artillery tactics in 1868 and 1869. The Barry Board's work included a Gatling gun drill, which was incorporated into the 1874 assimilated artillery manual.[40]

The Gatling gun raised questions about its organization and employment that the army debated for years.[41] Many officers were reluctant to accept the rapid-firing weapons. The most famous example occurred on the eve of the Battle of the Little Big Horn, when George Custer declined Alfred Terry's offer of a platoon of three Gatlings to accompany the Seventh Cavalry. Custer had a sensible reason for refusing the machine guns: mobility was vital in Indian warfare, and the Gatlings were ill suited to the rough terrain of the West. The guns that Custer left to Terry indeed proved a hindrance. Lieutenant James H. Bradley, who marched with the Montana column to the Little Big Horn, recorded that one evening "the cry did go up: 'The battery is missing!' A halt was made, and after some racing and hallooing the missing guns were set right again, having lost the human thread and so wandered a mile or so out of the way." Lieutenant Edward J. McClernand recalled "descending a long and precipitous hill, where it was necessary to fasten many lariats together, tie them to the Gatling gun carriages and then lower the latter by hand." McClernand also remembered that later during Terry's march, "the battery, especially, had great difficulty in keeping up. Several times it was lost and only brought back by repeated trumpet calls."[42]

The questions that the Gatling gun raised for the artillery were part of a larger dilemma. In the late 1860s and 1870s, rapid improvements in weapons had left every arm of the service in a quandary about its tactics. Thoughtful veterans remembered the hideous losses suffered by attacking infantrymen at Malvern Hill, Gettysburg, Cold Harbor, Franklin, and elsewhere and lost confidence in close-ordered assaults. Perceptive soldiers also foresaw that, with defenders gaining breechloaders and improved artillery, attacks would become even more dangerous in future combats than they had been on the dismal battlefields of the Civil War. One veteran of that conflict, Nelson A. Miles, had come up through the ranks to command a division during the Petersburg campaign, when rifle-and-trench warfare came to its culmination in the eastern theater. Miles wrote a letter in 1877 that envisioned future improvements in weapons and logistics but, significantly, said nothing about any corresponding progress in tactics. "In the next war," he predicted, "the breech-loading rifle will give place to the magazine gun, the old models of artillery will disappear and the science of equipping, supplying & muniting a command will be better understood." William T. Sherman was well aware that his army must be willing to change its tactics to keep up with improvements in weapons. He estimated that the American soldier of 1880 could deliver twelve times the firepower of his counterpart of 1779 and imagined that, "if Baron Steuben were to arise, he would doubtless attack one of Upton's thin lines with his old column of attack doubled on the center and would learn in a single lesson that the world has advanced in science, if not in patriotism, courage, and devotion to duty."[43]

It was far easier for the commanding general to declare that new tactics were needed than it was for his subordinates to agree on what those tactics should be. There were several reasons for this. One was that the army's thinking about strategy and tactics continued during the 1870s and 1880s to stress the importance of taking the offensive. Sir Edward Bruce Hamley's *Operations of War*, which Sherman praised as an excellent book,[44] became a text at West Point in 1870 and was taught at the Artillery School in the early 1870s and at the School of Application for Infantry and Cavalry, established at Fort Leavenworth, Kansas, in 1881. Hamley balanced his discussions of the advantages of offense and defense,[45] but his strongest words warned against fighting solely on the strategic and tactical defensive. "To pursue such a course, then," he admonished, "even when very inferior in force, is suicidal in a defender." American commanders continued to regard taking the offensive as a principle of war, and they recognized that, even though attacks had become more dangerous, some situations would require them.

"The fact that breech-loading rifles [have] made comparatively easy the defense of any building or intrenchment by a small force against a large one," Sherman noted in 1879, "does not alter the fact that such points must be defended, or must be carried[,] according to the object aimed at."[46]

A second issue was the practical matter of a small-unit commander's ability to control his men. Emory Upton made a case for his system of "fours" and single rank tactics, hoping that loose order and flexibility would help attackers regain parity with defenders. Until the invention of the field radio, however, an officer could not disperse his men far and still communicate orders to them. Alexander S. Webb, a brigade commander during the Battle of the Wilderness, calculated that if the Union force in that engagement had been "properly disposed for battle" in two ranks, with a third of its numbers in reserve, the Federals would have occupied a front of twenty-one miles. If nineteenth-century soldiers were deployed in loose formations across such distances, their officers could not control them. Upton himself acknowledged that "the safety of an army cannot be intrusted to men in open order with whom it is difficult to communicate."[47]

Still another problem was that some officers refused even to consider new tactics. At least one soldier concluded within four years of Appomattox that the army already had given too much study to the question. "After 'cramming' through Scott, Hardee, Casey and Upton," he said wearily, "we hoped that all would join in the exclamation, *Ohe! jam satis* (O! now, there is enough)." William H. Morris believed that there were officers who in 1866 still preferred Scott's old, Mexican War–era musket tactics to Hardee's rifle tactics or, for that matter, to any other system. Morris identified the opponents of tactical innovation as "the old prejudiced fogies" and those who wanted "to preserve the difficulties of the profession—as doctors stick to their prescriptions in latin and hyerogliphics." Another Federal veteran, T. C. H. Smith, believed the resistance came from "the bulk of the old officers aware of having to learn new things."[48]

While some officers resisted new tactics, many others ignored the subject altogether, once they had left West Point. During the late nineteenth century, military science became more complex, requiring soldiers to devote more time to staying current with professional developments, and many did not make the effort. Charles D. Parkhurst, who graduated with the Academy's class of 1868, worried in 1892 that a cadet "is filled with mathematics but not grand tactics. He learns the drill of the three arms of the service, but the drill only; study and labor for years after graduation, are necessary to keep him up to the progress made in the art of war shown by modern battles." During the early months of the Civil War, the volunteer soldier Jacob D. Cox had observed that most career army officers spent little time reading about tactical theory after they left West Point, and he asked Gordon Granger, a regular, about this. "'What would you expect,' [Granger] said in his sweeping way, 'of men who have had to spend their lives at a two-company post, where there was nothing to do when off duty but play draw-poker and drink whiskey at the sutler's shop?'" Cox believed his comrade's remarks were "picturesquely extravagant, but [they] hit the nail on the head, after all." The commanding general of the army, William T. Sherman, expressed concern that after young officers left West Point, they were not likely to give much more thought to the science of war. "My experience has been," he reflected in 1881, "that Graduates after leaving the Academy, if studiously inclined, are more apt to rest content with the knowledge [they have] already acquired of Tactics without further study, than of mathematics, mechanics, chemistry, geology, &c, which are more attractive." Four years later, one soldier disparagingly claimed that few captains could drill their companies properly and that still fewer officers knew how to drill a battalion.[49]

Those officers who were interested in studying tactics had trouble finding books on the subject. The soldiers assigned to the Department of Missouri during the 1870s and early 1880s could not have read very much about the science of war: their commander, Brigadier General John Pope, complained that the libraries on every post in his jurisdiction had "disappeared" during the Civil War and had not been replaced. The cadets at the United States Military Academy fared only somewhat better. "At the close of the [Civil War]," Jacob D. Cox wrote of West Point, "there was no instruction in strategy or grand tactics, in military history, or in what is called the Art of War. The little book by [Dennis Hart] Mahan on Out-post Duty was the only text-book in Theory, outside the Engineering proper." A cavalryman complained in 1873: "It is well known that we have had breech-loading arms for the last ten years, yet never a line from any competent authority as to how they shall be used." As for the third arm of the service, the artillery, Inspector General Randolph B. Marcy examined the textbooks available in 1880 and found only one that he considered satisfactory. He dismissed the others as, "for the most part, obsolete" because they included "nothing regarding the improvements and changes in guns, carriages, &c., for the past fourteen years."[50]

If a soldier managed to collect a few books on tactics, he probably had little chance to read them. The army of the late 1860s and 1870s, a small organization that had to contend with Reconstruction, Indian warfare, and daily routine, could not spare much time for professional studies. M. C. Meigs, a well-seasoned officer, acknowledged the press of day-to-day business when he advised in 1879 that it was more useful for a soldier to study "the duties of a Company officer" than "grand strategy." The commanding general of the army expressed his concern the same year that "less time is given to drills and professional instruction than should be the case." Colonel John Gibbon, a perceptive veteran of the Civil and Indian wars, toured the Department of Dakota at the time of these comments and heartily agreed with them. Dissatisfied with the drill and target practice of the troops he inspected, Gibbon blamed "the small size of the companies and the large drafts made upon them for working parties to keep up the ordinary routine labor of the posts." John Pope complained that a similar situation prevailed in his Department of the Missouri, where the "constant work imposed on" his infantrymen "both as laborers and soldiers in the field" left them with "little of the time possessed by [the] more favored arms of [the] service for drills or other military exercises." Pope advocated assemblying "a large number" of companies "for purposes of discipline and instruction."[51]

It would be unreasonable to expect the post–Civil War army to have arrived at what twentieth-century officers would consider a body of tactical doctrine. The nineteenth-century army had no permanent process for creating and revising an officially authorized set of tactical principles, recognized and taught throughout the service. Those few officers who were inclined, and were able to find the time, to study tactics could pursue the subject on their own, and they might—or might not—eventually get their ideas before a board of their peers.[52]

The small group of soldiers who evidenced an interest in tactics disagreed among themselves over several fundamental questions of theory, and the army entered the 1880s without a consensus about its tactics. In the late 1860s some officers began to envision grim days when breechloaders, Gatling guns, and improved artillery would create future battlefields even more dangerous than Antietam and Chickamauga. But other soldiers found inspiration in an earlier time, before the Civil War, when leveled bayonets and sharpened sabers had prevailed. These disagreements did not dismay the army's commanding general, who accepted them as part of the natural order of things. Writing in 1881, Sherman recalled the debates of the late 1860s and early 1870s and calmly reflected: "New arms, new habits, and new ideas were engendered by [the Civil War], and good men, skillful officers differed then as now and always."[53]

2

Hard and Dangerous Service
The Challenges of Indian Warfare, 1865–1890

For some officers the army's post–Civil War debates offered a diverting mental exercise but for other soldiers, stationed along the country's far frontier, tactics were a practical matter of life and death. Indian fighting intensified after Appomattox, when white migration into the West accelerated and Confederate defeat allowed the bluecoats to return their attention to western service. In 1865 the Indian wars remained an ongoing struggle, with murky origins in colonial times and no resolution in sight. Many years after the event, historians would use the Wounded Knee tragedy to mark the end of the frontier conflict, but contemporary soldiers and native Americans could not possibly hold such a perspective. Long after 1890, tensions ran high between the races and made demands on the army's attention. "It was hard for the army to realize," a cavalryman wrote in 1900, "that the days of Indian campaigning really had ended."[1]

Although some strategic and tactical principles emerged from the combats fought between Appomattox and Wounded Knee, the Indian wars were not a conventional conflict. One prominent historian characterized the United States Army of 1866–1890 as "not so much a little army as a big police force." His research showed that the army waged roughly a thousand combats with Indians between those two years, and the vast majority of these actions were minor skirmishes rather than major battles.[2] This same scholar counted more than two dozen small engagements that took place during the two-year Red River War alone. Another Indian wars historian, Don Rickey, noted: "Most of the encounters with hostiles, of the scouts and other troops campaigning in the field, came in the form of hit-and-run brushes and skirmishes."[3]

The participants themselves understood all too well that they were fighting a difficult, unconventional war. An infantry veteran of the conflict reminisced after the turn of the century: "You chased elusive Indians over routes of 'alkali, rock and sage'; they usually got away from you and all you got in return were the jeers of the fellows who didn't happen to be out on that trip. . . . That the service was hard, rough, disagreeable, and dangerous, was a matter of course, too well understood to need mention, cause comment, or even to justify entry of the fact on the monthly return." The commanding general was keenly aware that his army was being asked to fight an undeclared, unconventional war, a task that he considered entirely a hardship and without glory. General William T. Sherman noted in 1869: "While the nation at large is at peace, a state of *quasi* war has existed, and continues to exist, over one-half its extent, and the troops therein are exposed to labors, marches, fights, and dangers that amount to war." Lieutenant General Philip H. Sheridan resigned himself to the fact that some among the public would criticize the army, no matter how it conducted its Indian campaigns. William B. Hazen endured both Civil War and frontier service and concluded that the latter, "altho . . . creditable, has not much credit given for it." Brigadier General George Crook regarded the Indian wars, weighed against all other American conflicts, as "the most dangerous, the most thankless, and the most trying."[4]

Frontier combat was indeed dangerous and trying. The western Indians proved among the most ruthless opponents that the United States Army ever faced. Even allowing for white exaggerations of the savagery of their enemy, frontier warfare was starkly brutal. Colonel Henry B. Carrington's report of the Fetterman disaster near Fort Phil Kearney in

December 1866 included a ghastly accounting of the mutilated bodies found on the battlefield. A year and a half later, Brevet Major General Alfred H. Terry reported that "two men were killed by Indians 10 miles from Fort Buford; when found they were pinned to the ground by 27 arrows, scalped and horribly mutilated." Soldier-ethnologist John G. Bourke developed an open-minded interest in the Apaches, but he was not blind to their cruelties. Bourke described the "grewsome graveyard" that he found at Fort Bowie, "filled with such inscriptions as 'Killed by the Apaches,' 'Met his death at the hands of the Apaches,' 'Died of wounds inflicted by Apache Indians,' and at times 'Tortured and killed by Apaches.' One visit to that cemetery was warranted to furnish the most callous with nightmares for a month."[5]

Whites too were guilty of brutalities. The military tactic of encircling and storming Indian villages meant killing noncombatants: women, children, the infirm, and the elderly. General Crook accepted this outcome as inevitable during the course of his operations against Apaches in the Arizona Territory. He acknowledged that no matter what orders a commander gave before surprising an Indian camp, women and children were killed. After participating in an assault on an encampment of Warm Springs Apaches in 1882, one officer wrote home and blandly summarized the results for his mother: "We know of sixteen Indians we killed[,] twelve bucks[,] three sqaws and one child." In addition to these casualties of military actions, other Indians fell victim to frontier justice at the hands of civilians. Major General John M. Schofield affirmed in 1871 that some whites in Arizona were guilty of acts as barbarous as those they ascribed to the Apaches. Writing from the same territory two years later, George Crook alleged that "a floating class of miners" had carried out the "most horrible outrages" against the Walapais. In July 1887 the *Army and Navy Register* advised its readers that one Mexican governor had offered $700 each for Apache scalps. Many native Americans suffered because whites were not always willing or able to control their Indian allies. A stark entry appears for January 20, 1867, in William McKay's journal of the Snake War: "Met [our scouting party] coming back with 9 scalps had demolished and annihilated the camp & not one escaped then traveled 3 miles further killed 1 man 1 woman & child & surprised another camp and demolished it. Killed 5 women and 1 child."[6]

The Indian wars were a brutal conflict, and they came at a time when the army, severely reduced after the Civil War, was overburdened. General Schofield contended in 1885 that "the little Army of the United States is not and never has been on a 'peace footing.'" Brigadier General Oliver Otis Howard reported from the Department of the Columbia in 1880 that his troops were not only dealing with "wide–spread Indian discontent" but were also devoting their time to constructing and repairing post facilities, setting up telegraph lines, and building roads. In 1874 General Sheridan offered an even longer catalog of tasks that his men had been performing in the Military Division of the Missouri.[7] Beyond their responsibilities in the West, the bluecoats maintained a presence in the South until 1877.

Even when the army could narrow its attention to Indian fighting, it found that its mission was complicated by the difficulty of separating friendly Indians from combative ones. By the national government's own accounting in 1866, it considered "hostile" fewer than 100,000 Indians among a total population of 270,000. During the late 1860s military leaders found some of the Cheyenne and Arapaho on the Southern Plains to be cooperative and others perennially belligerent, and the same situation bedeviled the army during the Red River War of the next decade. No soldier was more frustrated by the problem of separating "friendlies" from "hostiles" than the Civil War hero Winfield Scott Hancock, who told a gathering of Kiowa chiefs at Fort Dodge in April 1867: "I want all friendly Indians south of the Arkansas to stay there, so that our young men won't mistake them for Sioux and Cheyennes."[8] During his tenure as commanding general of the army, Ulysses S. Grant expressed concern that "a good part of our difficulties arise from treating all Indians as hostile when any portion of them commit acts that makes a campaign against them necessary."[9]

Many senior officers hoped that this dilemma could be resolved by systematically confining tribes on clearly defined reservations. The rationality of this policy appealed to the military mind: Indians who remained on their assigned lands could be considered friendly, and those who left, hostile. In December 1868 William T. Sherman urged three of his subordinates to take action against the Cheyenne, Arapaho, and Kiowa bands, and then "mark out the spots where they must stay, and then systematize the whole, (friendly and hostile) into camps with a view to economical support, until we can try and get them to be self supporting like the Cherokees and Choctaws." Brigadier General John Pope summarized his interpretation of the reservation policy, in a report penned in 1870. "So long as the Indians remain on these reservations," he affirmed, "they are solely under the jurisdiction of the Interior Department. When they go beyond, it becomes the duty of the military to compel their return." Winfield Scott Hancock concluded in 1872: "the time has arrived when it is but merciful and just to the Indians as well as the whites to interfere with the nomadic habits of the former, . . . even should it be necessary violently to place them on reservations and rigidly to keep them there."[10]

Field commanders found, however, that forcing the tribes onto reservations, or indeed conducting any Indian operation, was a difficult business. Several factors hindered the army, beginning with the size of the regions where it had to campaign. "But tho' the country be large," one officer lamented, "the Army is small." Colonel Nelson A. Miles claimed that his troopers and mounted infantrymen covered more than a thousand miles during operations in the District of the Yellowstone in 1879. The Rio Grande country provided another extensive theater, where, John Pope complained, it was impossible to watch the long border and prevent raids across it. Colonel Benjamin H. Grierson decided that the problems of campaigning in west Texas could not "be conceived by any one unacquainted with the nature and extent of the country." Another veteran complained that, in the Southwest, the well-mounted Indians operated across a thousand miles of daunting terrain.[11]

The expanses of the West included some of the most difficult landscapes the United States Army ever confronted. "The country between the Yellowstone and the Missouri," Nelson A. Miles wrote, "is chiefly rolling prairie, occasionally interupted [*sic*] with tracts of 'mauvaises terres' or bad lands, from 10 to 20 miles in extent. There is very little timber, and in places a scarcity of water." The Southern Plains also posed obstacles to the army. Brevet Major General George A. Custer reported to his superior

Philip H. Sheridan that terrain had hindered operations after the Seventh Cavalry stormed Black Kettle's village along the Washita. Custer explained that he had not pursued the Indians after this November 1868 victory because his troopers had used up the provisions they carried, and ravines and other broken ground prevented his supply wagons from following him. Nowhere did the army encounter more rugged terrain than the deserts, mountains, and canyons of the Southwest. George Crook chased the Chiricahua Apaches along trails so treacherous that some of his pack mules could not keep their footing and were lost over the mountain precipices.[12] Crook's opponents used this unforgiving terrain to their best advantage. "The whole Sierra Madre," John G. Bourke declared, "is a natural fortress."[13]

Unrelenting weather punished the Indian-fighting army. The commander of Fort C. F. Smith, Luther P. Bradley, noted that the hottest day of the summer of 1867 fell during the month of the Hayfield and Wagon Box fights. The Civil War veteran recorded that August 20 featured "the most scorching sun I ever saw; nothing in *Georgia* or *Alabama* to equal it." George Crook described one region of western Arizona as a barren desert, where for half the year the temperatures registered 100 to 120° in the shade.[14]

Winter brought other forms of suffering. When, during the summer of 1876, Nelson A. Miles advocated a cold-weather campaign, William T. Sherman warned him that "winter on the Yellowstone is another matter from winter on [the] Red River." The commanding general reminded Miles that many hardy souls had died in the Montana winter, when the temperature might drop to forty below zero. Within a few months the field commander learned firsthand the truth of Sherman's counsel. "I had two enemies worse than Indians to contend with," Miles recounted in January 1877, "very severe elements and a terrible country to move over and the danger of getting snowed up without food was my greatest anxiety."[15]

The journalist De B. Randolph Keim left an account of life at Fort Cobb during the winter of 1868–1869, when the post was beleaguered by incessant rains. "Officers sat upon their bunks," Keim wrote, "in order to keep out of the wet. . . . The whole camp was thus subjected to a sort of water embargo, which was infinitely worse than snow or cold." George Custer reported of the Washita campaign: "This command was marched constantly five days amid terrible snowstorms, and over a rough country covered by more than twelve inches of snow. . . . The night preceding the attack, officers and men stood at their horses' heads for hours awaiting the moment of attack, this too when the temperature was far below the freezing point." Guy V. Henry, one of the most rugged veterans of the Indian wars, believed that "only a plainsman knows what it is to face a Norther. It is more dreadful than fire or shipwreck." Henry wrote with the benefit of his own cruel experience during a January 1874 march through the Black Hills. He returned to his quarters with his face blackened and swollen by frostbite, and each of his fingers frozen to the second joint, his flesh peeling off to the bones.[16]

Colonel Joseph J. Reynolds's men endured arctic temperatures during their March 1876 journey to the Powder River. John G. Bourke recalled that the expedition's supply of bacon froze so soundly that the cook had to chop it with an axe and that one morning the "thermometer failed to register. It did not mark below -22° Fahrenheit, and the mercury had passed down into the bulb and congealed into a solid button, showing that at least -39° had been reached."[17] Deploying to charge a village of about a hundred lodges along the Powder River,[18] Reynolds's troopers discarded some of their extra clothing and later suffered from frozen feet, fingers, ears, and noses. Reynolds still had not recovered from the ordeal of this expedition on April 11, when he sent a dispirited letter to the senior officer of the army. "General[,] these winter campaigns in this latitude should be prohibited," the field commander entreated from Fort David A. Russell. "Cruelty is no name for them—the month of March has told on me more than any five years of my life."[19]

Vast distances, rugged terrain, and crippling weather combined to create difficult, and sometimes impossible, problems of logistics. The January 1875 storm that struck the Red River force led by Nelson A. Miles offered one bitter example. Miles pointed out that a "Norther" typically spent itself after three days, but his expedition suffered bad luck that winter, when the storms proved unusually hardy. Episodes like this one helped convince Miles, whom the Sioux and Cheyenne called "Bear's Coat," that logistics were the key to all Indian operations. "The most important question," he wrote in March 1875, "is that of supplies—sources, routes, and means of transportation." That same winter, a battalion of the Ninth Cavalry scouted the Pease River country and found that although all but one of its horses survived, its mules could not. Major A. P. Morrow reported that fifty-four of his pack animals died in the heavy snow and rains. George Crook's men were reduced to eating horsemeat when their rations gave out during the Slim Buttes campaign of 1876. Stephen C. Mills, a veteran of the Twelfth Infantry, ridiculed the wagon-based logistics that his unit tried to use for the Bannock War. The Indians led the soldiers a merry chase through the difficult terrain of the Northwest, the regiment's supply system failed, and the infantrymen fell prey to an entrepreneur who had accompanied their expedition in his buggy. Mills gave a sarcastic edge to his reminiscences about this private contractor: "He was whole-souled, always happy, always had horses and whiskey sufficient to meet the demands and told me afterwards he made over $50,000 on his summer's contracts. So that hunting Indians, on strategic plans, with infantry in wagons, did win out for somebody."[20]

The Indian-fighting army also endured insects, dangerous mammals and reptiles, thirst, and the many other hardships of traveling in the nineteenth-century West. One veteran said of a barge trip on the Colorado River: "We tied up nights, slept and fought mosquitos about equal shares of the time." An officer stationed at Fort Bliss, Texas, during the late years of the Indian wars complained that the post was infested with rattlesnakes. Luther P. Bradley recorded these bleak notes in his journal on August 30, 1873: "We marched all day without water, and bivouacked at 8 [P.M.] on a bare hill side, no water or wood to be had, and no one Knows how far it is to the Yellowstone. Mules are giving out fast. 10 shot to-day, unable to travel any further."[21] An 1886 Apache campaign proved so strenuous that Captain Henry W. Lawton saw his weight drop by forty pounds, and his comrade Leonard Wood, then a young medical officer, lost thirty pounds.[22]

Beyond these hardships, the soldiers faced a dangerous, well-armed enemy. William T. Sherman acknowledged that his army admired the boldness of their Indian opponents. In 1911 Nelson A. Miles lavished praise on his former foes. "The celerity and secrecy of their movements," he wrote in his memoirs, "were never excelled by the warriors of any country. They had courage, skill, sagacity, endurance, fortitude, and self-sacrifice of a high order." Miles believed that native Americans fought best either in close combat, or when success buoyed their morale. "The Indian is a most dangerous warrior within two hundred yards," the veteran observed, "the range within which he is accustomed to kill game. . . . The Indian is also very brave—when he is successful." On another occasion Miles said of Sitting Bull's warriors: "I presume they would close in very readily if we gave them a chance, as there is no discount on their courage."[23]

On the Great Plains the army found the Lakota, called the Sioux—"the enemy"—by both native American and white opponents. This region was also the home of the Northern Cheyenne, whom John G. Bourke observed were well armed in the 1870s, carrying the latest magazine weapons and using metallic cartridges. Anson Mills praised the braves that he fought against at the Battle of the Rosebud: "These Indians lived with their horses, were . . . then the best cavalry in the world; their like will never be seen again." Major Marcus Reno's description of the final assaults against Reno-Benteen Hill at the close of the Battle of the Little Big Horn emphasized the determination of the Plains warriors. "In this attack they charged close enough to use their bows and arrows," Reno related, "and one man lying dead within our lines was touched by the coup-stick of one of the foremost Indians. When I say the stick was only about ten or twelve feet long, some idea of the desperate and reckless fighting of these people may be understood."[24]

On the Southern Plains the army fought against the redoubtable Comanche, Kiowa, Arapaho, and Southern Cheyenne. When criticizing or praising these warriors, soldiers consistently acknowledged their skill. George Custer revealed his own biases when he castigated the Southern Cheyenne as "the Dog Soldiers, the most mischevious, bloodthirsty and barbarous band of Indians that infest the Plains." In 1870 John Pope characterized the Kiowa as "altogether the worst Indians we have had to deal with." In contrast to these negative comments were those of one company officer who survived a sharp action against several hundred Kiowa and

Comanche attackers and praised them for staging "a wonderful display of horsemanship."[25]

This soldier's tribute was well placed, for the Plains tribes were renowned for their riding and for their ability to use either their mounts or terrain for cover. Edward S. Farrow, who successfully commanded a force of Indian scouts during the Sheepeater War, admired these skills in the Plains warrior. "He exhibits marvelous abilities in horsemanship," Farrow noted, "when fighting upon an open prairie he will frequently cast himself on the opposite side of his pony, until a foot on the back and a face under the neck of the pony are all that can be seen, and will fire with great accuracy either above or under him while at full speed." Anson Mills used much the same words to describe the riding and fighting techniques of his foes at the Rosebud, and Nelson A. Miles praised the ability of Sitting Bull's braves to protect themselves behind mounds or in ravines, presenting only their heads or puffs of smoke as targets.[26]

West of the Plains in the Rocky Mountain country, the army was opposed by the Utes, Paiutes, Bannocks, Snakes, and Nez Percés. William T. Sherman characterized the Utes as a "very warlike" tribe, well armed with "the best rifles and ammunition." Major General Henry W. Halleck, who commanded the Military Division of the Pacific during the late 1860s, spoke disparagingly of the Paiutes, Bannocks, and Snakes, but he grudgingly admitted they were "not without courage." The Nez Percés posed a particular challenge to the army: among all the North American tribes, they made the greatest use of conventional tactics. William T. Sherman credited their skill in deploying advance guards and rear guards, throwing out skirmish lines, and constructing fieldworks. Regular Army officers rarely acknowledged holding Indian warriors in much regard as strategists, but soon after Chief Joseph's surrender John Gibbon expressed hope that the leader of the Nez Percés would "give an account of his campaign, which was certainly very ably conducted."[27]

In the Northwest, the Modocs stoutly resisted the army from their stronghold in the lava beds of northern California. An ambitious young leader, Kintpuash or Captain Jack, was responsible for the death under a flag of truce of Brigadier General Edward S. Canby, the only general officer of regular rank killed during the Indian wars. This episode so infuriated the commanding general of the army that he was willing to accept the genocide of the Modocs. "All the Modocs are involved," William T. Sherman declared, "and do not pretend that the murder of General Canby was the individual act of Captain Jack. Therefore the order for attack is against the whole, and if all be swept from the face of the earth they themselves have invited it." Philip H. Sheridan largely agreed; his only regret about the Modoc War, seemingly, was that the Indians had not been exterminated. "I was very sorry to see the Modoc War end as it did," Sheridan lamented in June 1873. "[Brevet Major General Jefferson C.] Davis should have killed every Modoc before taking him[,] if possible."[28]

In the Southwest, the army contended against one of the most difficult opponents in its entire history, the Apaches. John G. Bourke, who knew these people as well as any officer did, wrote: "They carry almost nothing but arms and ammunition; they can live on the cactus; they can go more than forty-eight hours without water; they know every water-hole and every foot of ground in this vast extent of country; they have incredible powers of endurance; they run in small bands, scattering at the first indications of pursuit. What can the United States soldier, mounted on his heavy American horse, with the necessary forage, rations, and camp equipage, do as against this supple, untiring foe?" George Crook believed that the native American of the Southwest was a natural warrior. Revealing the biases of his time and place, the general compared his Apache opponents to wolves, coyotes, and tigers. Like many whites who characterized their Indian opponents as savage predators, Crook at the same time admired their courage and endurance. He reported after campaigning against the Chiracahuas: "They are not dependent upon the waterholes for water, but can go (100) one hundred miles without halting, carrying such water as they need for themselves in the entrails of cattle or horses killed by the way, and abandoning the animals they ride when these drop exhausted by thirst and fatigue."[29]

The best evidence of the fighting skill of the Indians was their success in combat. Their most famous battlefield victory, at the Little Big Horn in late June 1876, was but one episode of a campaign in which the braves of the Northern Plains prevailed over the army. The commanding general of the army found it galling that the native warriors had been able to defeat his field commanders. Sherman was particularly critical of George Crook's retreat from the Rosebud, which he believed had made possible the Indian victory at the Little Big Horn. "Surely in grand strategy we ought not to allow savages to beat us," Sherman complained to Sheridan, "but in this instance they did." The well-seasoned frontier veteran Charles King made this assessment: "[Crook] was up against an Indian proposition the like of which neither he nor Sheridan had ever known before."[30]

The Little Big Horn, or the Greasy Grass, as the Sioux called it, was not the only battlefield success won by native Americans between Appomattox and Wounded Knee. On December 21, 1866, north of Fort Phil Kearny, an Indian war party annihilated Captain William J. Fetterman's force of seventy-nine soldiers and two civilians, in less than an hour. "The scene of the action told its own story . . . ," wrote an officer who visited the battlefield the next day. "The arrows that were spent harmlessly from all directions show that the command was suddenly overwhelmed, surrounded and cut off while in retreat."[31] In January 1873 a few score Modocs successfully defended their stronghold in the California lava beds, driving off a mixed force of volunteers and regulars. White casualties came to nine killed and twenty-eight wounded, two of these mortally, but not a single Modoc was struck by military fire. In March 1876 Colonel Joseph J. Reynolds seized a Powder River village of Cheyenne and visiting Sioux families and captured its pony herd, but the Indians abruptly counterattacked, put the troopers into retreat, and regained their animals. The tenacity of the warriors proved a decisive factor in this back-and-forth battle.[32] At White Bird Canyon in June 1877, the Nez Percés had but three of their warriors wounded while slaying thirty-four soldiers.[33]

At the Rosebud on June 17, 1876, the Sioux and Northern Cheyenne won a strategic, if not a tactical, decision over a force of cavalry, infantry, and Crow and Shoshone auxiliaries, and stunned their commander, George Crook. The Plains warriors struck Crook so suddenly and so hard that the bluecoat commander lost both the tactical and strategic initiative and never regained either one. Assessing the operation in private correspondence, William T. Sherman called Crook's retreat from the battlefield of June 17 "a terrible mistake." The commanding general believed that if his subordinate had advanced instead of falling back after this engagement, "the Custer Massacre would have been an impossibility." The blows the Indians gave Crook along the Rosebud, and the later news from the Little Big Horn, turned the usually stolid frontier fighter into an uncertain commander who remained uneasy even after he had retreated to his base on Goose Creek. A month after his combat with the Sioux and Northern Cheyenne, the bewhiskered veteran continued to find things to worry about: that Colonel Wesley Merritt was taking too long to reinforce him, or that war parties might set fire to the parched grass around his camp. "I am in constant dread of an attack . . . ," Crook despaired on July 23. "I am at a loss what to do."[34]

The Sioux and Cheyenne gained their success at the Rosebud in part by catching their opponents off guard with an unusual display of conventional tactics. Lieutenant Colonel William B. Royall's report of the battle credited the Plains warriors with an ability to seize on tactical opportunities. Royall noted that at one point the Indians gained a "plunging and enfilading fire" against some of his men, that Captain Guy V. Henry's company had to check "a flank and rear movement of the enemy," and that later Royall's command was "subjected to a severe direct, flank, and rear fire." There is other evidence, beyond the example of the Rosebud, that native Americans had some appreciation of conventional tactics. Edward S. Farrow credited them with understanding the value of attacking against a flank. Chief American Horse and his Sioux followers fought from behind dirt breastworks at Slim Buttes,[35] and the Nez Percés were well known to employ conventional tactics.[36]

But these were the exceptions, not the rule. The Indians thrived on unconventional tactics, often making use of surprise, decoys, ambush, and evasion. "Strategy loses its advantages," Edward S. Farrow believed, "against an enemy who accepts few or none of the conventionalities of civilized warfare. The Indian is present one day and when next heard from is marauding in another state or territory; and oftener still, when supposed to be many miles distant, he is in ambush almost within range." Randolph B. Marcy, a veteran of the Civil and Indian wars, considered "the strategic science of civilized nations"[37] of little use when fighting the Indian, "an enemy who is here to-day and there to-morrow; . . . who is every where without being any where; . . . who comes into action only when it suits his purpose, and never without the advantage of numbers or position."[38]

Above all, the Indian proved an elusive foe. John G. Bourke compared pursuing an Apache band to swiping at fleas. Brevet Major General Orlando B. Willcox reported on a frustrating hunt for a Chiracahua band in 1882: "It was like a stag-chase in the hills without hounds." One long-time observer of the Plains Indians noted that the Cheyenne were sometimes called the Kite Indians because it seemed that they were always seen at a distance and while fleeing.[39] For decades many students of the Little Big Horn have believed that, in an ironic way, the elusiveness of the Sioux and Northern Cheyenne contributed to the defeat of the Seventh Cavalry. They have assumed that Custer attacked when he did, and divided his force, because he was concerned that the Indians would escape him.[40]

Elusive in maneuver and fierce in combat, the native American warriors waged courageous resistance for decades. Their opponents, small units of bluecoats stationed on the frontier, were called on to sustain an undeclared, ill-defined, and brutal conflict across vast expanses and in extreme weather. The Indian wars confronted the army with a challenging combination of strategic and tactical problems.

3

The Same Principle
as at Atlanta
Tactics and Strategy
of the Indian Wars,
1865–1890

American military leaders never prepared a formal statement of Indian-fighting doctrine. If the warfare on the
frontier had been a conventional conflict, with a well-fo-
cused mission for the army, its leaders might have paid
more attention to strategic and tactical theory, but the Indian wars did not follow that pattern. They were, instead,
a long-running police action, dating back to colonial
times, in which the army's task was broadly understood
but never precisely defined.

Historians have faulted the American military for its
lack of a formal body of tactical doctrine for western warfare. One prominent authority pointed out that Indian
fighting "was the army's primary employment for a century. Yet never did its leaders face up to the problem of
doctrine." Another historian largely agreed, contending
that the army's "organization and tactics were always directed toward possible wars with conventional, Euro-
pean-style military powers, and tactical manuals and
West Point teaching described this kind of warfare. Indian fighting, it was assumed, would soon be a thing of
the past; conventional warfare would become the army's

principal responsibility in the future." A third well-known scholar has argued that the general officers who were veterans of the Civil War, a large-scale conventional conflict, did not consider the fighting on the Plains to be a "real war" and that these leaders "who had commanded corps and divisions tended to be more interested in refighting Gettysburg or Five Forks in their memories than in formulating new strategies of unconventional warfare to deal with their Indian foe."[1]

While it is true that the army developed no formal tactical doctrine for Indian fighting, two other points should be kept in mind. One is that while the War Department did not authorize any manual on Indian warfare, in the way that it had endorsed Upton's tactics, a few soldiers studied the subject on their own initiative. One Indian wars scholar has identified two such officers, Edward S. Farrow and Randolph B. Marcy, and has called Farrow's privately published *Mountain Scouting: A Handbook for Officers and Soldiers on the Frontiers* the "closest thing to a manual on the subject of Indian fighting." This same historian called attention to two works by Randolph B. Marcy, his War Department–authorized *Prairie Traveler*, which gave some attention to Indian trailing and fighting, and his personal memoirs, *Thirty Years of Army Life on the Border*, which dealt more extensively with these topics.[2]

The second point is that doctrine is not the same thing as theory, and the lack of one does not mean the absence of the other. At least one historian has pointed out that "doctrine," in the context of military operations, is a twentieth-century term and has cautioned against applying it to the nineteenth-century army.[3] Although the soldiers stationed in the West during the 1870s and 1880s had in their hands no formal doctrine authorized and published by the War Department, there is evidence that they recognized a body of theory, a collection of ideas about Indian-fighting strategy and tactics based on experience and common sense.

One example of this body of theory was the strategic concept of winter warfare. During this season of cold, snow, and ice, the native Americans lost the advantages of mobility, as their grass-fed ponies were weaker than the grain-fed horses of the cavalry, and of evasion, as the severe weather fixed them in their villages. Winter brought the soldiers their best opportunity to find and defeat the Indians, or if a band fled, to destroy its camp and supplies. Philip H. Sheridan always credited himself with introducing the winter-war strategy in 1868. Paul Hutton judiciously pointed to many precedents, dating back to the eighteenth century, while acknowledging that Sheridan may have been ignorant of them.[4] John Gibbon favored this idea, arguing that the bluecoats should strike "when their [the Sioux] movements are restricted, their watchfulness less efficient, and any 'signs' left in the snow as plainly read by a white man as by an Indian." John G. Bourke also understood the advantages of winter campaigning and noted how the northern storms weakened the Indians' mares and foals in particular. "They become very thin and weak," he pointed out, "and can hardly haul the 'travois' upon which the family supplies must be packed. Then is assuredly the time to strike." Winter war was usually associated with the Plains but also found application in the Southwest. The commander of the District of Arizona concluded in 1866 that pursuing an Apache band was useless: only a midwinter campaign against their rancherías and provisions would defeat them.[5]

Another strategic theory, sometimes used in conjunction with winter warfare, was the motif of columns converging from three or more directions on the same region, in an effort to entrap any hostiles within the area. The Southern Plains campaign of 1868–1869, in which Sheridan believed he had introduced winter warfare, featured this multiple-columns strategy. In that operation three separate contingents, commanded by Major Andrew W. ("Beans") Evans, Brevet Major General Eugene A. Carr, and Brevet Brigadier General Alfred Sully, accompanied by Sheridan himself, converged on the Canadian and Washita valleys.[6] The Red River War of 1874–1875 provided another example, when five columns participated in a campaign against Southern Cheyenne, Comanche, and Kiowa bands.[7] The most famous case came in 1876, when the strategy was carried out by three forces: one under George Crook, another under Alfred Terry and John Gibbon, and a third under George Custer.[8]

The converging-column, and other, offensive strategies depended on locating the Indians, itself a challenging task. William T. Sherman described the difficulty of pursuing the Kiowa and Comanche raiders in Texas during the early 1870s: "It is worse than looking for a needle in a hay stack, rather like looking for a flea in a large clover field." Writing at

Stephen C. Mills of the Twelfth Infantry with Chiricahua scouts at Fort Apache, Arizona Territory, in 1881. To conduct offensive operations against Indians, the army first had to *find* them. Soldiers often relied on native auxiliaries like these to locate enemy camps. (Photograph courtesy of United States Army Military History Institute)

about the same time, Colonel Ranald S. Mackenzie complained that his Indian opponents rarely allowed him a chance to strike them. No warriors proved more elusive than the Apaches. "They take to the mountain ranges," one frustrated lieutenant colonel wrote in 1881, "where unseen behind rocks they pour a murderous fire upon their exposed pursuers led by some gallant young officer indescretely [*sic*] but eagerly charging the few who purposely show themselves—or they flee with the speed of the antelope if the force opposed to them is large or reinforcements come up, leaving our clumsily mounted troops far behind." George Crook concluded that the only way to pursue a band of Chiricahuas was to find their trail and follow it relentlessly.[9]

Crook further believed that this tracking could best be done by native

Americans. His early thinking about the use of Indian scouts and auxiliaries was founded on his own biases and questionable assumptions. "Nothing makes these 'Bucks' feel so good, as the idea of their being part of our
soldiers," Crook claimed in 1871, "& nothing will demoralize the hostiles so
much as to know that their own people a[re] fighting in the opposite ranks."
His presumptions aside, the bewhiskered Ohioan well deserves the credit
he has been given for championing this innovation. "I always try to get
Indian scouts," he told one journalist in 1876, "because scouting is with
them the business of their lives. They learn all the signs of a trail as a child
learns the alphabet, it becomes an instinct." Seven years later, Crook elabo-
rated his ideas about Apache warfare: "In operating against them the only
hope of success lies in . . . using their own methods, and their own people
with a mixed command. The first great difficulty to be met is to locate them,
and this must be done by Indians scouts." The use of native videttes and
allies had precedents in the nineteenth century—during the Second Seminole War, the army employed Seminole and other Indian guides[10] and had
an auxiliary Creek regiment—and even earlier, in colonial warfare. Yet
many officers of the post–Civil War army, ignorant of these earlier experiences, credited Crook with pioneering the use of Indian scouts.[11]

The army's use of native guides and auxiliaries was not limited to the
Apache campaigns. Crow and Shoshone allies accompanied Crook to the
Rosebud, where they raised the alarm of, and then helped fend off, the first
attacks of the Sioux and Northern Cheyenne.[12] John Gibbon detailed six
Crow volunteers to scout with Mitch Boyer for George Custer during the
Little Big Horn campaign. Lieutenant James H. Bradley rated these half-
dozen trackers among Gibbon's best guides.[13] Indian auxiliaries, including
a party of Pawnees under Frank J. North, made up about a third of the force
that Ranald S. Mackenzie sent against Dull Knife's village in November
1876.[14]

Some soldiers criticized their Indian allies. "I had no confidence in the
Indians with me," Marcus Reno declared after the Little Big Horn, "and
could not get them to do anything." Winfield Scott Hancock noted that in
his Department of the Dakota in 1869 fewer than half the authorizations
for Indian scouts were filled, and he acknowledged that some commanders
remained reluctant to use native guides. Two years later George Crook

complained that the army as a whole was using only a quarter of its allot-
ment of such billets. Peter T. Swaine of the Fifteenth Infantry, who observed the operations of two companies of Apache scouts, made this assessment in 1881: "I do not consider them reliable as soldiers for they have
nothing to gain by seeking a fight and it is rather to their advantage to lose
a trail than to find or follow one." Captain Guido Ilges of the Fourteenth
Infantry was disgruntled by the performance of his Pima and Maricopa
allies during an April 1867 operation in the Arizona Territory. "They en-
camped and marched contrary to their usual customs," he complained,
"and contrary to all common sense; exposing themselves to any lurking
scout, and lighting fires in the most exposed positions, both by day and
night." Over the course of the Indian wars, native scouts turned on their
white comrades only once, at Cibicu Creek in August 1881,[15] but this soli-
tary episode reinforced the distrust of Apache allies that many military
men harbored.[16] Philip H. Sheridan counted himself among those who
believed that Indian scouts might try to capture other Indians, or convince
them to surrender, but they could not be relied on to kill them in combat.
Sheridan, as commanding general of the army, and Secretary of War
William C. Endicott in March 1887 opposed George Crook's publishing his
arguments in favor of using Indian scouts.[17]

It is significant that within a month Secretary Endicott changed his
mind about this issue, for many officers believed that native allies were a
necessity. Their best-known advocate, George Crook, contended in 1886
that there had never been any successful operations against the Apaches
without the help of friendly Indians. "The longer we knew the Apache
scouts," John G. Bourke reflected, "the better we liked them." Bourke ap-
plauded Crook for having the wisdom to fight "the devil with fire." Henry
W. Halleck claimed that officers in Arizona in 1868 unanimously endorsed the value of these scouts.[18]

Beyond the Southwest, commanders on every sector of the frontier used
and praised Indian allies. Brevet Major General C. C. Augur touted Frank
North's Pawnee scouts in 1867: "They are unequalled as riders, know the
country thoroughly, are hardly ever sick, and never desert." Two years
later Augur reported that his Pawnees, operating from Fort McPherson,
proved indispensable in finding and following trails. Nelson A. Miles com-

mended his thirty Sioux and Cheyenne allies for their valuable service and brave fighting during the pursuit of Chief Joseph. On another occasion Miles urged that Indian or other "daring scouts" be used "to discover the hostile camps, trails or movements of Indians and the Cavalry saved for the direct march, resistless dash, and rapid pursuit, for which that arm of the service is so well adapted." At least one officer of the Tenth Cavalry, a black regiment, believed that native Americans made as good or better soldiers than whites or Negroes did. After nine years working with Indian scouts at several different posts, Lieutenant R. H. Pratt concluded there was "no good reason why Indians should not enter largely and permanently into Army organization, and all my experience fully sustains this opinion." Pratt reported in October 1876: "Without much effort I have taught my organization [of Indians] here [Saint Augustine, Florida] many of the simple movements of tactics. . . . Their comprehension and execution is so good that I feel safe in saying I can take a full company of fresh plains Indians and in six months drill put it through all the movements of a company, and conform to those of a regiment." Perhaps the strongest endorsement of the use of native Americans was that, as one historian of the Indian wars put it, department commanders "bombarded" the commanding general and the adjutant general with requests for permission to raise more Indian scouts.[19]

Another widely accepted concept, this one a matter of tactics, was the importance of deploying advanced guards, flankers, and rear guards to protect a unit and its supply train against surprise attacks, a specialty of native warriors. When the Seventh Cavalry marched up the Arkansas River in 1868, Edward S. Godfrey pointed out that the regiment moved with two troops in the advance, two on each flank, and two as a rear guard. Godfrey also described the traveling formation the unit used during its far more famous Little Big Horn campaign. "After the first day," he recalled, "the following was the habitual order of march: one battalion was advance guard, one was rear guard, and one marched on each flank of the train. . . . The battalions on the flanks were kept within five hundred yards of the trail and not to get more than half a mile in advance or rear of the train, and to avoid dismounting any oftener than necessary." A journalist who traveled with General Sheridan's column in November

1868 reported that it "moved . . . with flankers and videttes, who kept a vigilant look-out for lurking war-parties."[20]

A march by a company of the Fifth Infantry and some troopers of the Sixth Cavalry through the Washita country in 1874 provided an example of how flankers or a rear guard might protect their parent unit and its trains. Captain Wyllys Lyman reported that when a band of Kiowa and Comanche warriors attacked the moving column, the commander of the rear guard reacted quickly. Lieutenant Granville Lewis "shifted his line to meet [the attack] to the right and rear of the train and opened fire. The enemy swerved around the rear of the train, and accordingly Lieutenant Lewis removed his men across to cover it." While Lewis took a nasty wound from a shot through his left knee, his rear guard skirmishers beat off the assailants. Lyman and Lewis won brevets for this successful action.[21]

An episode from the Modoc War illustrated the opposite case: failure to deploy flankers could prove costly. As Brigadier General James T. Kerr told the story years later, a captain of the Twelfth Infantry led a reconnaissance by about ninety officers, men, and Indian scouts, and when he halted his party at midday "permitted the advance guard and flankers to be drawn in." Kerr was uncertain whether the company officer knew about the mistake or not, but the general was quite sure about the outcome. Kerr related: "Nearly the entire force was grouped about in an open space having lunch when suddenly it was fired on by Indians concealed in the rocks." The captain and three other officers were killed, a surgeon was mortally wounded, thirteen men were killed and sixteen were wounded.[22]

A related tactical theory called for the use of skirmishers, spreading out infantrymen or dismounted troopers in loose order. The army did most of its Indian fighting on foot, and this formation allowed soldiers to take advantage of cover and required no fixed distances between them, only that each man remained in sight of his comrade to the left and right. The Battle of the Rosebud offered some examples of skirmish-order fighting. Each of the five companies that composed Major Alexander Chambers's battalion made good use of the formation.[23] Lieutenant Colonel William B. Royall reported how four troops of his cavalry, hard-pressed by the Sioux and Cheyenne, had fought desperately from a dismounted skirmish line to protect their horseholders and gain a retreat.[24] Years after the Battle of Snake Mountain, Frank U. Robinson recalled that a troop of the Second Cavalry had employed the formation when it stormed an Indian settlement. "We then doubled our pace," he related, "and rushed into the village in close skirmish order. . . . It was almost a perfect surprise." George Crook described a small but brutal action that took place in Oregon in 1857: "We killed a great many, and after the main fight was over, we hunted some reserved ground that we knew had Indians hidden. By deploying as skirmishers, and shooting them as they broke cover, we got them."[25]

There were a number of reasons why frontier commanders turned to loose-order tactics. Many of them were Civil War veterans and knew that the use of skirmishers had increased over the course of that conflict, in response to the deadly volleys of rifled shoulder arms. Indian fighters faced a firepower dilemma of their own, for the Western tribes had significantly better weapons after 1865 than before 1861. George Crook declared in 1883 that the breechloader and metallic cartridges had "changed the entire nature of Indian warfare. The Indians are now no longer our inferiors in equipment; their weapons of even ten years ago have given place to breech-loading arms of the best makers." The flexibility of skirmish order proved useful, too, given the rough terrain of the West and the informality of frontier combat. John Gibbon believed that the nature of the fighting in the West promoted the army's transition from the close order, touch-of-the-elbow tactics of Hardee and Casey to looser formations. "The peculiar drill of men in masses," he wrote in 1879, "and the 'elbow touch' of the regular soldier, admirable as they are in ordinary warfare, are utterly thrown away in contests with the Indian." Charles King left an account of the Battle of Slim Buttes which suggested how loose-order tactics were well suited to the rugged ground and the individualism of frontier warfare. King related that Alexander Chambers's "plucky infantrymen" climbed up a cliff in skirmish order and then changed front to their right, executing the movement "practically, not tactically."[26]

Flexibility was a factor, too, when troops carried out the converging-column motif on the tactical level, attacking an Indian village on two or more sides simultaneously. This tactic proved most successful when it was executed at dawn, as soon as the soldiers could see well enough to take their assigned positions, yet still so early that a camp had not stirred to life. Such attacks from two or more directions were employed or attempted in many battles of the Indian wars, including the Washita (November 1868), Stronghold (January 1873), Powder River (March 1876),[27] Slim Buttes (September 1876), Big Hole (August 1877), and Bear Paw Mountain (September 1877). Ranald S. Mackenzie wanted to encircle Dull Knife's village along the Red Fork in November 1876, but the canyon terrain allowed only one access to the encampment, through a gap from the east.[28] The aggressive George Custer probably intended to attack from two or more sides the enormous village that he found along the Little Big Horn. Custer had a strong precedent for separating his regiment. "The division of the [Seventh Cavalry at the Little Big Horn] was not in itself faulty," the veteran Edward S. Godfrey believed. "The same tactics were pursued at the battle of [the] Washita and were successful."[29]

The encirclement tactic, and other offensive operations, were appropriate tasks for the cavalry, the most mobile of the three arms of the service. When Indians attacked the Union Pacific Railroad during the spring of 1867, William T. Sherman advised Ulysses S. Grant that infantry could neither prevent these forays nor pursue the raiders and that only cavalry could deal with the contingency. An infantry major stationed at Fort Dodge that summer explained to his superior that he needed troopers rather than foot soldiers because the Plains warriors would not fight against infantrymen but instead "easily escape from them." In July 1876 Thomas L. Rosser, a veteran of the Confederate mounted service, reviewed the recent Little Big Horn campaign and noted that cavalry was the only arm suited to such an operation. Infantry that tried to follow war parties, he believed, "upon a mission such as Custer's . . . are as useless as fox hounds in pursuit of wild geese." The cavalry's operations during the two years after the Little Big Horn, Edward J. McClernand contended, proved that the mounted service could attack and defeat Indians. The ambitious Nelson A. Miles was the most vocal advocate of sending foot soldiers rather than horsemen against the western warriors, but even he recognized that at times cavalry was needed. "If I had had a regt or battallion of mounted men," he moaned on October 23, 1876, two days after his infantry won a victory at Cedar Creek, "I could have captured the entire body [of Sitting Bull's band] estimated at upwards of 400 lodges but it is not easy for Infy to catch them although I believe we can whip them every time."[30]

Miles persistently argued that the army had underestimated the value of infantry in Indian operations, a contention that was substantiated by the performance of the foot soldiers at Cedar Creek. Miles had put his thesis in front of William T. Sherman ten days before the Little Big Horn: "I do not believe in magnified scouts but believe that a good command at least one half Infantry can make [the Sioux] country so uncomfortable for them and giving them no rest they would be compelled to sue for peace." Sherman saw some merit in Miles's point of view. Faced with congressional cuts in the army's budget in 1878, the commanding general preferred reducing either the artillery or cavalry rather than the foot troops. "I assure you," he wrote to Major General John M. Schofield, "for fighting Indians, the infantry has eclipsed the cavalry, whose horses are usually played out, or are in the way at the moment of conflict, which must always be at the option of the Indians." Philip H. Sheridan agreed that by the late 1870s the cavalryman had become essentially an infantryman who used his mount for transportation, not combat.[31]

Discussions of the third arm of the service, the artillery, focused on how it could best be used on the offensive, since native Americans, unlike conventional troops, would not attack fieldpieces. The Gatling gun offered a dramatic example: it could deliver destructive fire against Indian warriors, but they seldom were so accommodating as to attack the rapid-firing weapon. During the Indian wars the field guns lost their defensive role, which they had played so well throughout the Civil War, and the service also suffered because so few of its contingents came from artillery regiments. The distinguished Union artillerist Henry J. Hunt was disgusted to see Gatling guns served in the late 1870s by "uninstructed cavalry and infantry officers and men, . . . with chance horses from the quartermaster's corrals, . . . and this while five dismounted light batteries were lying idle, so far as their proper duties were concerned."[32]

Among the three arms of the service, the artillery was the least suited to Indian campaigning. Hunt concluded that light batteries at "extreme frontier stations" were "of no possible use." Nelson A. Miles reviewed the fieldpieces available shortly after the Little Big Horn and criticized all of them, including Richard J. Gatling's weapon. "I am not surprised that poor Custer declined taking a battery of gattling guns," Miles reflected. "They are worthless for Indian fighting." He went on to suggest: "What we want is a gun similar to the light French steel gun or the one used in the Ashantis campaign of the English."[33]

Oliver O. Howard's experience at the Battle of the Clearwater in July 1877 suggested some of the difficulties of artillery in Indian operations. Approaching the Nez Percé village, the one-armed Civil War veteran fired at the encampment with a gun from his howitzer battery and two Gatlings,[34] but he used them at such long ranges that he succeeded only in giving away his position. Howard tried to employ his artillery during the two-day battle that ensued, but the rugged Idaho terrain limited its effectiveness. He acknowledged that his Gatlings made no contribution other than speeding the enemy's retreat from the battlefield.[35]

Soldiers learned that artillery, foreign to the experience of native Americans, could be used to intimidate them. One case in point took place in Arizona in 1867, when an Apache band annoyed a camp from an elevation above it. A lieutenant became impatient with the harassing Indians and turned a howitzer on them. The shell burst above the warriors, another officer recalled, "and as if by magic every Indian disappeared as if driven into the ground, and from that time forward not an Indian was seen or heard of in that vicinity. I never knew if any or how many were killed or hurt, but I feel certain that these Indians were greatly amazed, if not hurt, by this to them new system of warfare." Nelson A. Miles claimed in his memoirs that "Indians could not stand artillery." At the Wolf Mountains on January 8, 1877, Miles surprised a Sioux and Cheyenne war party by disguising two artillery pieces as wagons and then, at an opportune moment, removing their canvas coverings and turning them on his startled opponents.[36]

On other occasions, the artillery's effect was both psychological and physical. Lieutenant Colonel Thomas H. Neill of the Sixth Cavalry reported that he used a Gatling gun "very successfully" at four hundred yards, during an April 1875 action against 150 Cheyenne warriors. Nelson A. Miles employed both infantry and artillery to defeat a band of Sitting

Bull's followers in an October 1876 encounter. Four Hotchkiss guns contributed to bringing the Indian wars to their grim conclusion at Wounded Knee in December 1890, raking that ill-fated Sioux encampment from end to end. Having silenced the village, the soldiers turned the Hotchkiss cannon against a nearby gully where many of the Indians had fled. In the final, ghastly scene of the Indian wars, exploding artillery shells wreaked carnage and death among the fugitives in the ravine at Wounded Knee.[37]

Although most discussion of the artillery and the other two arms focused on their use on offense, it must be acknowledged that defense was an element in the army's strategy and tactics during the Indian wars. At the level of strategy, the bluecoats built and defended lines of posts, such as those along the Platte River and the Bozeman Trail and those across Kansas[38] and Texas.[39] At the tactical level, soldiers or frontiersmen were sometimes forced to fight on the defensive, usually successfully—as at the Hayfield and Wagon Box fights (August 1867), Beecher's Island (September 1868), Adobe Walls (June 1874), Reno-Benteen Hill at the Little Big Horn, or Milk Creek (September 1879).[40] On two dramatic occasions, bluecoat forces were put on the defensive and annihilated: this was the fate of Fetterman and his men along Massacre Ridge near Fort Phil Kearny and of Custer and his five troops on the hills east of the Greasy Grass.

For the most part, however, Indian operations meant offensive strategy and tactics: finding the tribal bands, overcoming their mobility and elusiveness, and defeating them in battle or forcing them onto reservations. Historian Robert M. Utley described the army's Indian wars strategy as one of "total war." Russell F. Weigley made the same assessment and titled the chapter of *The American Way of War* that dealt with the Indian wars "Annihilation of a People: The Indian Fighters." Robert A. Wooster later rejected the idea that military leaders arrived at a consensus for a total-war or war-of-annihilation strategy,[41] but he agreed with these other two historians that the army's strategic thinking stressed taking the initiative. Wooster described the strategy of the Indian wars as moving through a series of phases, with an emphasis on offense alternating with one on defense and eventually prevailing over it. In his summary of events on the frontier after 1865, the army attempted to protect settlers and railroads with a defensive strategy immediately after the Civil War, then assumed the offensive in 1867 and 1868, had its aggressiveness dampened by the peace policy of the Grant administration's first term, and finally "took the field with a vengeance" until major Indian resistance was crushed.[42]

Indian-fighting officers repeatedly urged offensive strategy and tactics, none more staunchly than Nelson A. Miles. Reflecting on his preparations for the Red River War, Miles said he "resolved upon certain principles" that he "regarded as essential," among them, "always [be] ready to act on the offensive." The "Bear's Coat" wrote of the Plains Indians: "If the offensive is assumed and persistently maintained, even by inferior numbers, they are sure to give way,"[43] and, on another occasion, "by constantly acting on the offensive I found that they could be discouraged and dispersed."[44]

Other officers heartily advocated aggressive strategy and tactics; they often aimed their comments at the Apaches in particular. Henry W. Halleck contended in 1866 that these Southwestern Indians "must be hunted and exterminated." A year later Brevet Major General Irvin McDowell wrote: "They can only be successfully fought by troops who carry on an *offensive* warfare against them who do not wait until they have attacked, . . . but who fight them in their own way." Ulysses S. Grant accepted an Indian peace policy when he was president, but as commanding general he had minced no words about the Apaches. "There is no alternative," he wrote in 1867, "but active and vigorous war till they are completely destroyed, or forced to surrender as prisoners of war." "If there is a good chance," Ranald S. Mackenzie asked his superior officer early in June 1873, "please let me attack the Apaches." Later the same month, Mackenzie pressed his request: "I believe that if the Mescaleros or Lepans could be hit pretty hard that the Kickapoos would conclude to come out."[45] In 1885 Philip Sheridan, like Grant eighteen years earlier, advocated attacking the Apaches until they were killed or captured.[46]

Commanders commonly urged offensive operations against other tribes. After the Fetterman disaster in December 1866 Colonel Henry B. Carrington advocated taking the initiative against the raiders of the Bozeman Trail. "They cannot be whipped, or punished," he contended, "by some little dash after a handfull, nor by mere resistance of offensive movements." William T. Sherman gave a ringing endorsement to offensive operations in an 1868 letter to three of his field commanders. "I am well satisfied with Custer's attack [at the Washita]," he declared, "and would not have wept if he could have served Satanta's and Bull Bear's bands in the same style. I want you all to go ahead, kill and punish the hostile, rescue the captive white women and children, capture and destroy the ponies, lances, carbines &c, &c, of the Cheyennes, Arapahoes, and Kiowas." The autumn after Crazy Horse contributed to the Indian victory at the Little Big Horn, George Crook called for a relentless offensive against the Cheyenne leader. "[Crazy Horse] should be followed up and struck as soon as possible," Crook urged. "There should be no stopping for this or that thing." The peppery Philip H. Sheridan became impatient with the defensive strategy of building a line of forts across Texas because, as he was quick to point out, the Indians would eventually avoid the posts and find new routes for their raids.[47] Sheridan consistently favored aggressive operations[48] and charged John Pope in 1874 with advocating defensive ones. Pope's distraught reply revealed his belief that offensive strategy and tactics were axiomatic among frontier officers. His distressed, long-winded defense included these sentences: "That an officer of my long experience in the Indian Country & experience among Indians should advocate defensive instead of offensive operations, is simply incredible unless such a want of common sense is also charged, as I may safely say, I have not generally been charged with. As a mere question between offensive & defensive operations against Indians there can be no dispute among men who know anything."[49]

Carrington, Sherman, Crook, Sheridan, Pope and many other officers of the Indian-fighting army were Union veterans of the Civil War. This enormous conflict was the central event of their careers—indeed, for most veterans, of their lifetimes. The experience of the Civil War lingered with these men into the Indian wars, if sometimes only in nostalgic or ephemeral ways. George Custer compared his attack at the Washita, about three and a half years after Appomattox, to those he had led against the Confederates. "The moment the charge was ordered," he wrote, "the band struck up 'Garry Owen' and with cheers that strongly reminded me of scenes during the war, every trooper, led by his officer, rushed toward the Village." Custer offered a more specific Civil War analogy when he described the battle to Philip H. Sheridan, his commander during the Virginia and Shenandoah Valley campaigns. Referring to the destruction of a Confederate corps during the Appomattox campaign, the dashing cavalryman told his superior that the Washita had been "a regular Indian 'Sailor's Creek.'"[50] Guy V. Henry remarked to a journalist that the 1879 Milk River battlefield reminded him of Petersburg, where he had commanded a brigade. Luther P. Bradley, who had been wounded at Chickamauga, carefully noted the anniversary of that terrible battle in a journal he maintained in 1867 and 1868 while serving an isolated frontier assignment. Nelson A. Miles was amused to hear his troops sing "Marching Through Georgia" during the Red River War.[51]

These nostalgic memories that Civil War veterans took with them into the West carried no weighty consequences: far more important was the influence that the campaigns of the 1860s had on the strategic thinking of the leaders of the Indian-fighting army, most of whom had been Union senior commanders. Robert M. Utley and Russell F. Weigley concluded that these officers formulated an Indian wars strategy derived from the offensive, total warfare they had waged in both theaters of the Civil War in 1864 and 1865. "Sherman and Sheridan were of a single mind on strategy," Utley believed. "Atlanta and the Shenandoah Valley furnished the precedents. Like Georgians and Virginians four years earlier, the Cheyennes and Arapahoes would suffer total war." Discussing Sheridan's 1868 planning for winter operations, Weigley wrote: "The strategy Sheridan chose was an innovative one for an Indian campaign, reflecting his and Sherman's experience in carrying war to the enemy's resources and people."[52]

In *The Military and United States Indian Policy, 1865–1903*, Robert Wooster offered a different point of view, arguing that the origins of the army's Indian wars strategy were largely independent of its Civil War experience. Stressing the differences between the two conflicts,[53] Wooster concluded that the leaders of the Indian-fighting army, in contrast to the senior Union commanders, did not reach a consensus in favor of pursuing total war.[54] He believed that the strategy of the western army, rather than being a unified one derived from Civil War practice, was the product of a haphazard mix of the experiences of individual commanders and their responses to immediate circumstances, the influences of army politics, and the conditions of frontier warfare.[55]

Wooster did acknowledge one point that Utley and Weigley stressed: during both the Civil and Indian wars, army commanders emphasized an offensive strategy.[56] Virtually all of the senior leaders of the Indian-fighting army were veterans of the decisively victorious Union armies. Many of the subordinates of Grant and Sherman in the Atlanta, Carolinas, Shenandoah Valley, or Virginia campaigns of 1864 and 1865 later saw frontier service: Philip Sheridan, George Crook, Ranald S. Mackenzie, Winfield Scott Hancock, Nelson A. Miles, E. O. C. Ord, Oliver O. Howard, John M. Schofield, George Custer, Alfred H. Terry, and John Gibbon. Their Civil War experience persuaded these veterans that the North had won by seizing the strategic offensive in May 1864, in both theaters of the war, and relentlessly maintaining it to Appomattox and Durham Station. Sustaining the strategic offensive had meant, for the most part, maintaining the tactical initiative as well.[57]

This experience produced an Indian-fighting army whose senior leaders strongly believed that offensive strategy and tactics won decisive results. They urged taking the initiative, using phrases that recalled their Civil War experience. While Indian-campaign plans were being considered during the autumn of 1868, Grant asked Sherman: "Is it not advisable to push after their villages and families? The possession of them would bring them to terms." Answering criticism of assaults on Indian encampments, Sheridan pointed to Civil War precedents: "During the war did any one hesitate to attack a village or town occupied by the enemy because women or children were within its limits? Did we cease to throw shells into Vicksburg or Atlanta because women and children were there?" At the opening of 1878 the ambitious Nelson A. Miles lobbied for authority to undertake an independent offensive campaign against the Sioux. "Now suppose you suggest to the Secty of War—or President," he pressed the commanding general, "to give me one chance at this business alone. It can not be managed by men a thousand miles away any more than you could have won the Atlanta campaign by remaining in Nashville." In 1867 Sherman contended that "defensive measures" to protect the Union Pacific railroad against Indian raiders would prove useless and that offensive operations were necessary. "I want to attack them," he declared, "where they are known to be in force, viz., on the Powder and Yellowstone.... It involves the same principle exactly as when I was moving on Atlanta. If I had turned back as the Rebels wanted, and as my own people to the rear clamored for, we would now be down in Georgia scrambling for the safety of our grub."[58]

During the tactical debates of 1865–1880, many officers had maintained a confident optimism about offensive operations. Throughout the same years and later, into the 1880s, the army contended with the hardships and dangers of frontier fighting. Its leaders believed this conflict called for active campaigning, and the premise seemed ratified as their soldiers eventually subdued the western warriors. The Indian wars strengthened the army's confidence in the virtues of offensive strategy and tactics.

4

Individual Skill and the Hazard of Battle
Marksmanship and Training, 1880–1898

While the bluecoats waged Indian warfare in the West, some soldiers gave thought to the tactics needed for combat against a conventional foe. Much of this tactical thinking during the 1880s was closely related to the army's emphasis on marksmanship, as many American officers concluded that attackers could use accurate individual fire and loose formations to counter the improved weapons available to defenders. Major General John Pope, an old soldier of the Mexican, Civil, and Indian wars, predicted in 1883: "As the military inventions and conditions tend more and more to a system of skirmish fighting, which shall supplant the old systems of attack in line or in column, the value of the skilled shooting force will be more and more conspicuous." A company-grade officer asserted the same year: "Modern arms, and especially the conditions which govern war in this country, make skirmishing the most important item in the American soldier's tactical education." In 1886 Major General Alfred Terry noted that while in the past the most important difference between well-trained and raw units had been their ability to maneuver, now it rested in the skill of their individual soldiers with their firearms. Colonel John C. Kelton impatiently dismissed the argument that soldiers at small posts did not have time to improve their marksmanship and praised the example of a one-company station where, in spite of an ongoing construction project, the commander in 1881 had almost three-fourths of his men "constantly at drill and target practice."[1]

While marksmanship gained value with the introduction of improved weapons and loose formations, its significance had also been underscored during the army's frontier experience. General Pope characterized the Indian wars as a series of running skirmishes, in which skillful sharpshooting proved invaluable. Pope's fellow veteran of the Civil and Indian wars, John Gibbon, regarded good marksmanship as the first requirement of successful frontier fighting. Another of the most famous western soldiers, Edward S. Godfrey of the Seventh Cavalry, wrote an 1896 essay on the importance of fire discipline and marksmanship in dismounted cavalry combats against Indians and other opponents. A soldier who served farther down in the ranks from these officers explained in high-blown prose that the Indian wars had forced his unit to take its marksmanship training against armed warriors, rather than paper objects. "We have had living targets which have engaged our very thorough attention too often," this infantryman wrote in 1882, "and necessity has compelled us to neglect the target practice, just as it has prevented us from 'capering nimbly in a lady's chamber to the lascivious pleasings of the lute.'"[2]

While the army's experience during the Indian wars highlighted the importance of marksmanship, William Conant Church, editor of the _Army and Navy Journal_, and George W. Wingate, a National Guard officer, also did much to promote it. In 1871 Church asked Wingate to write a series of articles on rifle practice for the _Journal_, which Church published in that forum and also, four years later, under the title _Manual for Rifle Practice_. The two men took the lead in establishing the National Rifle Association, which was founded in 1871 and opened its own target range, at Creedmoor, New York, two years later.[3] Regular army teams competed against militia marksmen in the annual matches at Creedmoor during the 1870s and eventually won the match in 1880.[4]

The emphasis on marksmanship reached its zenith during the 1880s, a decade when the army, as Lieutenant Matthew F. Steele put it, "mounted" the hobby-horse of target practice, "a beast of long life and great endurance." By 1888 another of the army's intellectual junior officers, Eben Swift, could claim that the United States Army had developed the world's finest marksmen. The marksmanship craze of the 1880s led one infantry officer to recall the 1870s with nostalgia: "Those were the good old days.... Target practice was practically unknown. I think the allowance of ammunition was twenty rounds a year, and by custom of the service, it went to hunting."[5]

During the 1880s soldiers spent more time on the rifle range, and their target practice also became more sophisticated. In 1885 commanding general Philip Sheridan gave his infantrymen more realistic targets, profiles of soldiers representing an enemy skirmish line,[6] and commanders tried to improve the marksmanship of their men from loose formations. One officer explained their motivation: "It now [1880] being conceded that open order must be the rule in all future fighting, . . . increased pains with the individual instruction of the soldier in marksmanship becomes necessary." The Union Civil War hero Major General Winfield Scott Hancock strongly endorsed this idea, and Colonel Robert P. Hughes, a devoted student of tactics, urged in 1888 that "the nearer this practice firing as skirmishers is made to simulate" actual combat, "the more valuable it will be."[7]

Hughes had his soldiers practice on open, mown-grass fields where they fired at targets at fixed distances, and he also deployed his men in woods and on rough terrain, challenging them to estimate their ranges. While serving as inspector general in 1888, Hughes concluded that Captain Stanhope E. Blunt's *Instructions in Rifle and Carbine Firing* had helped the army improve its marksmanship but that its troops needed better training at estimating ranges. He firmly believed that a contest between "two equally good marksmen . . . one of whom has thoroughly learned to estimate distances and the other has not, would be about similar to a battle between two pugilists one of whom had one of his arms tied." Major Robert H. Hall of the Twenty-second Infantry was of the same mind as Hughes, having urged as early as 1886 that soldiers should practice skirmishing more often at unknown ranges and less frequently at known distances. In 1897 Lieutenant Colonel Henry C. Hasbrouck, a Civil War veteran, made the same sugges-

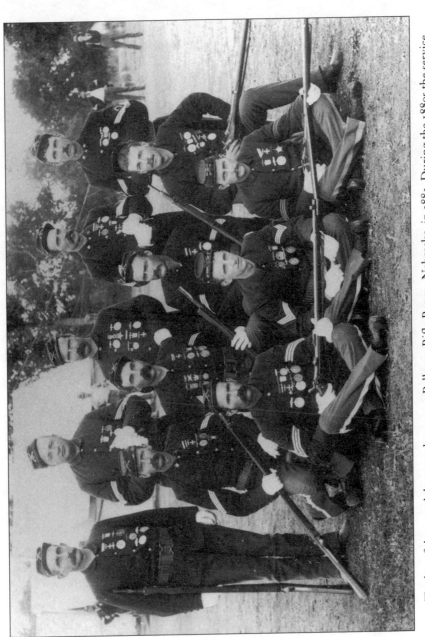

Twelve of the army's best marksmen, at Bellevue Rifle Range, Nebraska, in 1881. During the 1880s the service became infatuated with target shooting. (Photograph courtesy of United States Army Military History Institute)

PHILIP H. SHERIDAN.

Commanding general of the army during the height of the
marksmanship craze, Sheridan introduced a set of realistic targets:
profiles of enemy soldiers deployed along a skirmish line.
(Photograph courtesy of United States Army Military History Institute)

tion for artillery exercises: "The captains should not know beforehand the targets to be fired at. . . . The difficulty in field practice is to determine quickly the right elevation and length of fuse."[8]

Closely related to estimating ranges was practicing against moving, as well as fixed, targets. Hughes again was among the champions of this reform, arguing that both attackers and defenders should train to contend against a mobile opponent. Improved weaponry had forced infantry to advance in rushes, and therefore defenders would have to deliver accurate fire at a moving foe. Assaulting troops would need the same skill, and Hughes feared that American soldiers lacked it. "Actual tests show such an unreadiness in our men to fire quickly," he warned, "that the party on the defensive can rise from cover, deliver a volley, and drop back under cover before a single shot is fired at them from any part of the attacking line. This can best be remedied by practicing the troops in firing at moving and disappearing targets, which I think should be done." An 1883 _Army and Navy Journal_ article recommended that soldiers practice against progressively more difficult targets: figures of men standing, then kneeling, lying down, "and, finally, the targets should be fired at when in motion." Cavalrymen and artillerymen also needed to train for combat against a mobile foe. Captain Moses Harris of the First Cavalry urged in 1882 that troopers be required to fire at a target that would run out on a track, stand visible for about five seconds, and then be pulled back out of sight. Lieutenant John W. Ruckman of the First Artillery, who once called gunnery practice "the branch of our work which must not fail," proposed that crews should fire, day and night, against targets moving at varying speeds.[9]

The 1880s emphasis on target practice was part of a larger effort to use realistic practice in the field, rather than studying tactical theory or practicing drill, as the primary means of preparing for future combat. "The men must be taught the noise and confusion of battle," an officer of the Fourth Artillery urged in 1890, "in every way that is possible. . . . Do we ever seek out the roughest ground available and teach the men there?" Two years later Brigadier General August V. Kautz contended that most drill instruction proved impractical in combat. Drilling and target shooting, another soldier insisted, did not "go far enough." Young officers needed "[i]ncessant field work . . . [and] thorough criticism of this work by experienced superior officers." As early as 1879 the army's commanding general lectured a group of fledgling soldiers on the importance of practical training. William T. Sherman warned his eager young audience against believing "that when you have thus assembled your men, grouped them into companies and regiments, ranged in line of battle, equipped with, bayonets, knapsacks, haversacks, &c.,—every man and officer in the place prescribed by tactics[,] that you have made them *soldiers*. . . . I have known many an officer who knew Jomini by heart, and could demonstrate every battle of Frederick the Great, and of Napoleon, on the blackboard[,] who could not properly station a picket-guard, or handle a company skirmish-line, or know how to provide for his men on a ten days' scout."[10]

By the 1890s the army was pursuing vigorously Sherman's recommendation that more time be given to practical field exercises and maneuvers rather than parade-ground drills or classroom studies of tactical theory. In 1891 Colonel Edwin F. Townsend reported proudly that the Infantry and Cavalry School's Department of Military Art had participated in twenty-three field exercises during the past year, while "theoretical instruction was confined to 'Operations of war,' by Hamley." Taking advantage of the unusually good weather during the spring of 1891, Lieutenant Colonel Hamilton S. Hawkins, West Point's commandant of cadets, directed his young charges away from their classroom desks and ordered them outside to practice advance guard, rear guard, outpost, and reconnaissance operations.[11] A Civil War veteran who staunchly believed in practical training, Hawkins later served as commandant of cadets at the Infantry and Cavalry School, where he dismissed the student recitations of drill regulations as "a waste of time." To facilitate field practices at Fort Leavenworth, he removed the barbed-wire fences around the post and built crossings over its railways and ravines. Under the leadership of officers like Hawkins, the Infantry and Cavalry School abandoned its earlier emphasis on drill, ceremonies, and garrison duties.[12]

During the late 1880s and into the 1890s, army stations all over the country conducted tactical exercises. "In the past few years," an officer of Sixth Infantry wrote in 1889, "it has been the custom of our Service, wherever practical, to send troops on practice marches and remain in summer camps for short periods." The commanding general proudly reported in 1896:

"Most of the infantry, cavalry, and light artillery have engaged in practice marches and field maneuvers during the year, which have been highly beneficial to the service and instructive to both officers and men." Brigadier General John J. Coppinger believed that the battle exercises conducted in his Department of the Platte during 1895 represented an improvement over earlier close-order drills and ceremonies. That summer the secretary of war visited one of Coppinger's posts, Fort Robinson, Nebraska, where he saw eight troops of the Ninth Cavalry carry out "some interesting practical exercises in open order drill."[13]

The new field exercises were sometimes conducted as part of a post's inspection: the visiting inspectors described to the battalion or troop commanders a theoretical situation, such as an enemy advance against their station, and the officers deployed and maneuvered their men to meet this imaginary contingency. These exercises tested the ability of commanders to solve practical tactical problems and of their soldiers to operate in the field. Fort Crook, Nebraska, was "defended" by one battalion against another in 1897, while at Fort Clark, Texas, two cavalry troops and four infantry companies protected an entrenched camp against a comparable force, presumed to be advancing from the Rio Grande through Dolores.[14]

Army inspectors were keenly aware that soldiers needed realistic exercises as well as drill. Colonel Robert P. Hughes contended in 1890: "Parade-ground drill does not qualify the officers and non-commissioned officers for the work expected of them when they come in contact with a vigilant enemy." "The great need of the army," the inspector general urged in 1894, "is not theoretical instruction or routine duties, but the frequent and thorough application of principles, gathered from all sources, in the solution of tactical and other military problems." A year later one of his subordinates became dismayed when he inspected troops that had not trained under field conditions. "They are proficient in ceremonies, routine garrison duties, such as police, guard, etc., and in that portion of the drill regulations that relates to exercises in close order," he complained, "but in the extended-order drill over unknown country and on varied ground, such as they would be compelled to operate over in service, they are entirely without instruction."[15]

The inspectors expected to see realistic field exercises, and sometimes they saw greater realism than was safe for the participating soldiers. Cavalryman Alonzo Gray recalled his experience with a department inspector who, as part of his annual visit, dictated a field problem in which Gray's comrades were to stage a mock attack on some infantrymen escorting a wagon train. The captain leading the charge failed to control his men and, as Gray remembered, "the whole affair was a mess. The Cavalry charged through the train." The horses dashed down on the hapless foot soldiers, who scrambled for cover among the wagons, and many "were hurt. Lieutenant [Ernest Bertrand] Gose, who was on a horse, was bowled over and quite badly injured." The inspector wrote a "scathing report" of the fiasco but, Gray concluded, the troopers shrugged off the episode: "What did a Cavalry private care, since he 'could not control his horse.'"[16] Realistic field exercises could be hazardous for senior officers as well as their subordinates, as evidenced by a harsh accident that befell Brigadier General Guy V. Henry in July 1895. A veteran of the Civil and Indian wars who had been wounded in the face by a .44-caliber bullet during the Battle of the Rosebud, Henry found that peacetime operations could also be hazardous. While commanding a force in a mock battle at a New York National Guard camp, he rode across a soft hayfield and was severely injured when his horse stepped in a hole, turned a full somersault, and rolled over him. "The General's face was scratched," a newsman reported, "some teeth had been knocked out, and he said he thought his nose was broken."[17] Training exercises sometimes proved dangerous for the men in the ranks, too, as in the case of two Florida soldiers who were wounded during a June 1884 sham battle. The powder in the blank cartridges issued to these state troops was so old and dried that it had caked into small fragments, which struck human flesh with the same force as a lead bullet.[18]

The emphasis during the 1880s and 1890s on rigorous field exercises, realistic target practice, and skirmish formations, was necessitated by improved weapons and put an increased premium on the individual skills and initiative of both officers and enlisted men. "Modern war calls for a larger measure of intelligence on the part of the individual officer and soldier than it did twenty years ago . . . ," William T. Sherman observed in 1879. "Tactical changes are forced on us, . . . thus necessitating more study and preparation on the part of the professional soldier, than in former times."

Sherman knew from his own Civil War experience that during that conflict, successful combat leadership by lieutenants and captains had depended largely on an officer's personal bravery and his ability to keep his men in close-ordered lines. During the next war, a company officer would again be expected to set a courageous example and also, with his men deployed in skirmish formations and making attacks in short rushes, he would have to make quick decisions under fire, independent of his battalion commander. In 1894 Captain James S. Pettit predicted that in future combats all officers, from second lieutenants through senior generals, would have to act vigorously and show sound judgment. A company officer of the Thirteenth Infantry predicted in 1895: "The captain . . . will fight the battles of the future." Two years later, General Coppinger recommended that officers should be challenged to solve tactical problems "based upon an assumed situation, according to the leaders' judgment of its requirements, and untrammeled by hard and fast rules, . . . accustoming them to rapid decision and action." Secretary of War Redfield Proctor believed the army should reward self-reliant officers, young men who demonstrated "good sense."[19]

The tactical theory of the period challenged small-unit commanders to take more initiative. Emory Upton had been well aware of this development, and the tactics that he was preparing at his death called for junior officers to assume greater responsibilities. In 1891 the army adopted a new tactical manual, which expected captains to make a number of independent decisions, demanded by the stress of battle. "Each captain in the fighting line regulates the march of the line within the limits assigned him," it provided, "determines the distances to be passed over in rushes, and brings his support from the firing line, pursuant to orders, or without orders, if the necessities of the movement require; he directs the fire and regulates its intensity." The same volume assigned broad responsibilities to a battalion's major, who was expected to exercise "a general control, and [to endeavor] constantly to increase the energy of the action." At the Infantry and Cavalry School during the early 1890s, student officers were encouraged to write their own infantry and cavalry textbooks.[20]

Some theorists foresaw that, for better or worse, noncommissioned officers would have greater influence over the outcome of future battles. In 1889

Lieutenant H. J. Reilly proposed a modification of Upton's system of fours, in which each block of fours would serve as a "fire unit" or "group," under the "direct control and responsibility" of an NCO. Colonel H. C. Merriam of the Seventh Infantry endorsed a similar system, based on squads of eight soldiers, each consisting of a corporal who would control the seven privates around him, keeping them in formation. Merriam probably made this suggestion after studying the 1891 manual, which made a squad of two "fours," a corporal and seven privates, "the basis" of its loose-order tactics. Shortly after this system was adopted, a cavalry lieutenant commented on how it had increased the NCO's influence in combat, but unlike Reilly and Merriam, this trooper did not embrace the prospect. "In this all important matter of firing," P. D. Lochridge pointed out, "a captain used to be able to control his company. Now, a corporal may ruin it." By the mid-1890s the *Army and Navy Register*, which had initially endorsed the 1891 tactics with enthusiasm, worried that the new system put too much reliance on "the authority of the non-commissioned officers . . . under conditions which might call for special skill and judgement."[21]

Many officers realized, too, that this increased responsibility would fall not only to NCOs but also to all of the men in the ranks. H. G. Litchfield, a Civil War veteran, acknowledged in 1880 that a soldier could not "be in all cases fully instructed beforehand as to his method of fighting, and to his individual skill . . . much of the hazard of battle must be entrusted." Colonel Alexander McCook, commandant of the Infantry and Cavalry School, characterized "the individual instruction of the soldier" in 1887 as "now so important." Brigadier General Stephen Vincent Benét, long the army's chief of ordnance, offered the opinion late in the 1880s that the "days for shoulder to shoulder formations for fighting are passed, and Tactics must yield to the necessities of getting from our breech-loaders and magazine guns all the possibilities of the arm. This calls for the individuality of the soldier." Colonel Henry W. Closson explained how this aspect of the new tactics contrasted with the old. Parade grounds in the past had been used "to teach one man how to handle many," he reflected in 1891. "Their use now is to teach every man how to handle himself."[22]

Nor was the infantry the only arm that valued self-reliant leaders and practical field exercises. "Young graduates of West Point," a lieutenant of the Third Artillery declared in 1884, "need practice more than anything else." This officer also contended that every argument made "in favor of small-arm target practice and battle firing . . . proves still more strongly the necessity that exists for artillery target and battle practice." "It is not at all necessary that every officer in the service should be a *scientific artillery expert*," wrote another soldier in 1894, "but it is necessary that all should be *practical artillerists*." In 1888 First Lieutenant Tasker H. Bliss, who would later become chief of staff of the army, noted that several artillery-post commanders in the Division of the Atlantic had complained about the theoretical ideas the army taught its gun crews. Bliss proposed to answer their criticisms with field exercises, including "practical construction of batteries—field and siege," and "Distant Reconnaissances," which would challenge an artillery-post commander to plan an extended march from his station. Arriving on unfamiliar ground, his officers would have to study the terrain and decide where their batteries should be placed. Colonel Robert P. Hughes urged the same year that the artillery should make practice marches, "full of instruction for the younger officers and the men. . . . Each march might present its own peculiar lesson in artillery works: the choosing of positions, the lines of approach, the attack of positions, the attack or defense of defiles, estimating distances, etc."[23]

The artillery manuals of the 1890s also recommended practical exercises and encouraged independent action. The 1891 regulations recognized that battery officers would have to rely on their own judgment more often in future combats, acknowledging that in "field practice and actual service" commanders would not always be able to move their guns in "exact conformity" with the formal guidance in their official tactics. "In such cases," the manual advised, "the methods prescribed in the Drill Regulations should be regarded as types to be followed as closely as possible, each captain conducting his battery by the simplest means and shortest practicable route to the nearest available place for it in the new formation." The 1896 regulations, similar to 1891, gave the same guidance.[24]

The official manuals and many artillery officers recommended practical exercises, and the army put these proposals into practice during the 1880s and 1890s. The *Army and Navy Register* told its readers in April 1882 that instruction at the Artillery School gave "precedence . . . to the practical work of the artillerist." An observer of the target practice at the school that summer was pleased to see that "our modern artillerists are keeping up with the spirit of the age" and to report that the gunners made use of a number of innovations, including a field telephone. In 1892 Lieutenant Edgar Russel enthusiastically described a rigorous field exercise that Battery F of the Third Artillery had conducted from Fort Sam Houston, Texas, the preceding summer.[25]

The cavalry's equivalent of these artillery exercises were its rides of instruction, intended to give troopers practical experience in the field and to encourage officers to become more self-reliant. In 1887 Lieutenant John P. Wisser of the First Artillery published some examples of field exercises for all three arms of the service, including a ride of instruction for the cavalry, in which the troopers would practice their reconnaissance skills. Two years later a sizable force, including two cavalry regiments, practiced operating in the Indian Territory. Major John Breckinridge Babcock, a long-tenured cavalry veteran, reported enthusiastically on this field exercise and became a zealous proponent of the rides of instruction.[26] In 1894 Babcock sent Captain Eben Swift, Arthur L. Wagner's reform-minded assistant at the Infantry and Cavalry School's Department of Military Art,[27] a detailed proposal for an exercise at Fort Leavenworth. Babcock advocated a ride of instruction that would take the officers out of the classroom, confront them with theoretical problems in the field, and challenge them to make quick decisions on the spot. He urged that such field training would make each student "think for himself and . . . would make him *feel* for the time the responsibilities of the commanding officer, called upon to decide quickly and at the same time wisely, upon scant information of the enemy's strength and intentions." Soon after Babcock's proposal, the students began making reconnaissance rides from Fort Leavenworth, preparing more than three hundred maps of the roads around the post and using a field case expressly designed for carrying sketching equipment on these exercises.[28] Across the country at Fort Myer, Virginia, where Guy V. Henry commanded four troops, this energetic career officer also was sending his men on practice marches and map exercises during the mid-1890s. "These rides and reports of same should be required by the regulations and made uniform throughout the service," Henry declared, "and thus [would] bring

GUY V. HENRY.

A rugged veteran of the Civil and Indian wars, Henry made shrewd
observations about the late nineteenth-century army.
(Photograph courtesy of United States Army Military History Institute)

better results, more interest, and a knowledge of a horse's power not shown now by daily routine drills." A Civil War veteran, Henry also took his troopers to the Bull Run and Antietam battlefields, where his officers discussed the terrain with their men. Late in the 1890s, Lieutenant F. S. Foltz of the First Cavalry welcomed the opportunity to conduct rides of instruction and field maneuvers. Earlier, during the Indian wars, he reflected, "our time was only sufficient to rest the men and condition the animals for another expedition when the grass should be up in the spring."[29]

Like their cavalry comrades, engineering officers devoted more time to practical exercises during the 1880s and 1890s. Fieldworks had proven valuable during the Civil War, and although they had been far less common during the Indian wars, the army sustained its interest in fortifications, confident that they would prove important in any conflict against a European opponent. "War now, more than ever before," artillerist John C. Tidball declared in 1887, "demands the employment of field entrenchments, and too high an estimate can not be given to practical skill in such work." A few years before Tidball advanced this contention, Francis V. Greene, an engineer officer who was a close student of military affairs, argued that fieldworks rather than breechloaders were responsible for the dominance of the tactical defense over offense. Greene reasoned that in open fighting, attackers armed with breechloaders could load and fire while advancing, nullifying any advantage their opponents might have in rate of fire. It was only when the defenders protected themselves with entrenchments that they gained a decisive edge. Consistent with the urgings of the army's commanding general, William T. Sherman, who had seen the value of fortifications at Vicksburg, Atlanta, and elsewhere during the Civil War, the Fort Leavenworth curriculum encouraged students to develop their engineering skills in the field. Colonel Alexander McCook reported from the Infantry and Cavalry School in 1886: "A course of practical engineering was marked out, the officers making gabions and fascines, with all kinds of revetments known to modern engineering." By the end of the decade, the wide-ranging engineering work at Fort Leavenworth included: "construction of shelter pits and trenches, gun pits and epaulements, trestles and spar bridges to include spans of 45 feet on a scale of 1 inch to 1 foot, cask piers, gabions, f[a]scines, and hurdles and wire entanglement, defilading, profiling according to the latest type of hasty field works on a level, on slopes, and the use of blocks and falls, and cord-age."[30]

During the 1880s and 1890s, American officers gave serious thought to the principles of field entrenching and to the tactics of every arm of the service. Historian David Chandler, an authority on the Napoleonic era, once cataloged the distinguishing features of the British army at the end of the eighteenth century: "Rigid linear tactics, draconian discipline based on the fear of the lash, endless drill, and implicit discouragement of individual initiative—these were the most obvious characteristics of the red-coated soldiery of John Bull's island."[31] For most of the nineteenth century, the American army also was marked by three of these same attributes: inflex-ible linear tactics, tedious drill, and a stifling of individual initiative. Dur-ing the 1880s and 1890s, the United States Army abandoned these tradi-tional elements and began to move its tactics toward the twentieth century.

5

The Deadly Ground
Issues in Tactics,
1880–1898

Throughout the 1880s the United States Army sharpened its marksmanship, and during that same decade and the next, it improved its training. Yet it still faced what one military writer termed, as late as 1896, "the problem of the day." He referred to the continuing dominance of the tactical defense, which had emerged during the Civil War and had grown stronger during the decades afterward. Improved firepower created much of the dilemma. "Infantry fire against masses is now as effective at 2,000, 2,500, and even 3,000 yards, as formerly at 800 yards . . . ," Lieutenant Colonel Henry Lazelle counseled in 1881. "From five to ten shots may now be fired when one was formerly fired." Another careful student of military affairs, Captain Francis V. Greene, predicted in 1883 that magazine guns would increase this rate, even "in the excitement of battle," to fifteen shots a minute and the calmest soldiers might fire as many as thirty rounds a minute,[1] and machine guns would produce even more dramatic results.[2]

Other technological developments increased the rates of fire of weapons and improved their accuracy. Metallic cartridges became common after the Civil War, giving more efficient fire-power to both hand guns and shoulder arms. Breech-loading cannon re-placed muzzle-loading ones, increasing the firing rates of the artillery. This improvement served to reinforce the advantages already enjoyed by defending field guns. By the late 1880s, the European military powers were developing smokeless gunpowders, which would give defenders clear fields of fire against advancing soldiers.[3]

Successful assaults also would be more difficult in the future because engineers were designing increasingly sophisticated fieldworks. The army's schools devoted considerable attention to entrenchments during the 1880s and 1890s. During the mid-1880s students at the Infantry and Cavalry School as well as cadets at West Point were examined on Junius Brutus Wheeler's *The Elements of Field Fortifications*. Officers at the Artillery School were challenged with a "practical problem requiring each to prepare a plan for the city of Norfolk from an attack by land. . . . Each member of the class was required to prepare plans on a map of an entrenched line, giving the armament and number of troops required for each part and the reason for each operation." In 1884 one soldier praised the field engineering manual used at the Fort Leavenworth schools, claiming that it combined practical lessons from the Civil War with more recent advances.[4]

By the 1880s and 1890s professional soldiers were keenly aware of the daunting problem that sophisticated fieldworks and improved weapons posed for attackers. "Dash and enthusiasm," Captain James Chester declared in 1886, "are out of place in front of a line of breech-loaders covered by intrenchments." A few years earlier a British officer described the terrible challenge faced by advancing infantry: "A certain space of from 1,500 to 2,500 yards swept by fire, the intensity of which increases as troops approach the position from which that fire is delivered, has to be passed over. How shall it be crossed?"[5]

American soldiers sometimes referred to the ground in front of a defending line as the "danger zone" or "deadly zone." Major General Wesley Merritt used the phrase "danger space" to describe this area, where flat-trajectory fire would strike down many attackers. Writing in the mid-1880s, James Chester referred to the same ground as the "danger zone" or the "deadly space." In a later essay he distinguished between what he called a "danger zone" and an even more daunting region for attackers, a "deadly zone." The "danger zone" was the ground that began at the extreme range of the defenders' weapons, the outer area of combat where the advancing troops would first come under fire. The "deadly zone" was the terrain that began at the defenders' effective range, the terrifying stretch that the attackers must cross in their final assault. In 1889 Lieutenant H. J. Reilly grimly predicted that "the dangerous space" in front of a defensive position would become even more treacherous for attackers in the future.[6]

Once inside the "deadly ground," attackers would be able to execute only the simplest movements. The *Army and Navy Register* warned officers in 1891 that their troops would not be able to maneuver under the fire of contemporary weapons: they would be able only to move forward or backward. The veteran cavalryman Edward S. Godfrey agreed with this realistic point of view. "After getting within the limits of 'the rain of fire,' say 700 yards," he wrote in 1896, "it will be extremely difficult to move the firing line, except in advance or retreat." The officer who tried to maneuver or halt his troops within the range of "modern rapid-fire arms," another soldier explained, would pay a terrible price. Every "minute which a commander holds his troops under fire," this theorist warned, "on account of tactical evolutions, however graceful and beautiful, is an unnecessary loss of just so many lives."[7]

The firepower of entrenched defenders would also make it extremely difficult for attacking troops to maintain cohesion, regardless of their formation. This problem had hindered Civil War assaults, commonly made in close-ordered lines. Successive lines of advancing troops usually intermingled with one another and often became so hopelessly jumbled that even when an attack succeeded, the victors could not exploit their advantage.[8]

After the Civil War, continuing improvements in weapons intensified the problem. Lieutenant Colonel N. B. Sweitzer contended in 1884 that breechloader fire would cut up advancing lines and throw their companies into disorder. Sweitzer, a Civil War veteran, called on one of the tactical lessons of that conflict: "Experience has proved that it is impossible to avoid mixing men of different companies and of regiments when brigades are supported by brigades." A British essay, reprinted in the *Army and Navy Register* a few years after Sweitzer's comments, argued that attacking companies would inevitably become intermingled. Tactical theorists could no longer prescribe a distance between attacking lines with any confidence that their advice would prevent intermingling. "No nation experienced in war will undertake to say at [what] distance from the enemy the attack formation must begin," a British officer conceded in 1883, "nor the exact distance between the various portions into which the battalion is divided."[9]

By the 1880s most theorists agreed that frontal assaults by close-ordered lines moving at a cadence must give way to more flexible tactics. Arthur L. Wagner, in his publications and instruction at Fort Leavenworth, warned against traditional headlong advances. A thoughtful student of military history, Wagner emphasized the virtue of flexibility, stressing that attackers must be ready to make flank rather than frontal offensives, to make the best possible use of cover, and to take advantage of any weakness of a defender, such as low morale or faulty deployments. Some theorists concluded that assaults would have to be made by successive rushes rather than a sustained advance. Lieutenant Colonel Sweitzer made this assumption in his 1884 predictions about future infantry attacks. Captain James Chester agreed; he believed that the "deadly zone" would have to be "crossed with a rush." The official infantry tactics of 1891 were based on the same assumption: they provided for attacks by squads of soldiers in loose order, making short rushes from one covered point to the next.[10]

The field telephone had the potential to help attackers carry out these more flexible tactics. This new invention might allow commanders to communicate with dispersed troops and make loose-ordered formations a practical alternative to close-ordered lines. Before Alexander Graham Bell's device had any influence on battlefield tactics, it first appealed to soldiers because it was well suited to the rifle range. A field telephone provided communication between the firing line and target shelters during the Illinois National Guard's 1885 marksmanship practice. Two Signal Corps officers, Charles E. Kilbourne and Richard E. Thompson, designed a portable model that was used at the 1894 rifle competition of the De-

Signal Corps soldiers training with field telephones, about 1904. These instruments began appearing on target ranges during the 1880s, and technicians eventually improved them for service in combat. Through the Spanish–American War, however, officers still directed their men across the "deadly ground," as they had during the Civil War, by voice commands. (Photograph courtesy of United States Army Military History Institute)

partment of Texas. Brigadier General Frank Wheaton, a hardened field veteran, readily foresaw a battlefield application for the invention. If soldiers could use the portable telephone to advantage on their rifle ranges, he predicted they could also carry it on their skirmish lines, in their assaulting ranks, and with their storming columns.[11] The field telephone's descendant, the field radio, fulfilled General Wheaton's expectations and eventually helped restore the balance between attacking and defending troops. In the 1890s, however, the field telephone remained an emerging technology, not yet able to end the dominance of the tactical defensive.

Despite the supremacy of the defense, American officers continued to believe that offensive operations were essential, to keep the initiative and to gain decisive results. Faced with the enormous firepower in the hands of defenders, theorists might have compromised on this principle of the virtue of active warfare and advocated, as Confederate Lieutenant General James Longstreet had before the Gettysburg campaign, a combination of the strategic offensive and tactical defensive. This concept was sound in theory, but its best chances for success in the field depended either on an enemy foolish enough to attack to no purpose or on one who would allow himself to be outmaneuvered. If Maj. Gen. George G. Meade had permitted the Confederates to gain a crucial strategic position, one threatening Baltimore or Washington, for example, he would have been forced to attack them. But the Northern commander was not so obliging: he pursued a wise strategy and gave the Southerners no chance to implement Longstreet's theoretically sound idea. There was another limitation on the concept of taking the strategic offensive and the tactical defensive: even if a commander outmaneuvered his opponent, he would not necessarily be able to fight as a defender, a point that Captain John Bigelow made in his 1894 textbook on strategy. "It may be generally asserted," he explained, "that an army cannot move upon any commanding strategic point without overcoming some resistance, and that, once in the enemy's rear, it must move against him and attack him under penalty of being overpowered or seeing the value of its position nullified by the enemy's manoeuvres. *The strategic offensive ordinarily involves the tactical also.*"[12]

The United States Army's officer corps was not intimidated by this prospect. Many soldiers in both America and Europe continued to emphasize the virtues of seizing the tactical offensive, rather than the advantages of the defense, which the Civil War had underscored. This thinking was well represented in a recommendation made by the veteran cavalryman Edward J. McClernand, who in 1890 urged: "The battalion commander will avoid a passive defense by taking the offensive whenever practicable." Some officers, Americans as well as Europeans, believed that machine guns would help attacking troops as much as they would benefit defenders.[13]

The traditional faith in the bayonet also remained hale and hearty. A "grand bayonet charge" provided the capstone to a field maneuver conducted by the Rhode Island militia in 1882. Lieutenant John Bigelow, Jr., a military historian with a keen interest in tactics, praised the edged weapon in an essay the same year. Bigelow claimed that since the bayonet's introduction, no great victory had been won without it, and he predicted that it would prove of greater use in the near future than it had during the recent past. Another regular officer, William Powell of the Fourth Infantry, complained in 1884 that the edged weapon received too little attention. Some officers defended the bayonet on the grounds that, as Charles Parkhurst put it, "no matter what may be said as to the usefulness of the bayonet, its exercise doubtless teaches dexterity in the handling of the rifle."[14] The army's official tactical manual during the Spanish-American War devoted seven pages to a bayonet exercise.[15]

Musketry, the companion of the bayonet, continued to have its supporters. "Marksmanship has so monopolized the military mind these days," the conservative James Chester complained in 1891, "that musketry seems forgotten." Chester contended that the smoke of battle would prevent infantrymen from accurately estimating distances or taking aim, and if smokeless powder were used, the "excitement" of combat "would render their judgment worthless." After studying the European armies during the middle 1890s, some American theorists concluded that their own army should stop emphasizing individual marksmanship and adopt the German system of "field fire," in which the officers estimated the range and their men shot at their command. Matthew F. Steele reflected in 1895 that the army had "mounted" the hobby-horse of target practice, "a beast of long life and great endurance. At last, however, this good steed's back is swayed,

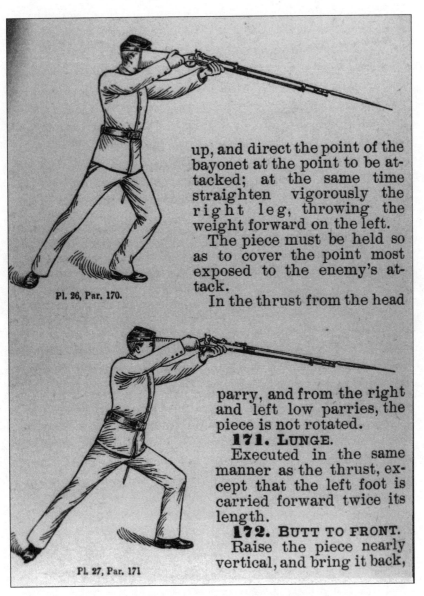

up, and direct the point of the bayonet at the point to be attacked; at the same time straighten vigorously the right leg, throwing the weight forward on the left.

The piece must be held so as to cover the point most exposed to the enemy's attack.

In the thrust from the head

Pl. 26, Par. 170.

Pl. 27, Par. 171

parry, and from the right and left low parries, the piece is not rotated.

171. LUNGE.

Executed in the same manner as the thrust, except that the left foot is carried forward twice its length.

172. BUTT TO FRONT.

Raise the piece nearly vertical, and bring it back,

Two plates from the 1891 infantry manual, illustrating its bayonet exercise. The new tactics introduced innovations but also retained some traditional ideas. The 1891 manual devoted seven pages to its bayonet exercise.

and he shows other signs of decrepitude, which makes us fear that he may break down under the weight of the Krag-Jorgensen, with its terrible length of range and heavy cost of cartridge." Cavalryman William H. Carter assessed the markmanship craze of the 1880s: "Target practice had, for a few years, been forced to the front to the exclusion of other instruction, but good judgement prevailed at last and it became an integral part of the whole, of recognized value, but not the only criterion of efficiency."[16]

While some infantrymen persisted in these traditional ideas, artillerymen suffered from an even more acute problem: the army neglected them. At the same time that foot soldiers argued about the merits of marksmanship and musketry, in 1888 the inspector general reported that some field batteries were conducting no practice firing whatever. In 1896 an officer of the Fourth Artillery complained that the artillery had failed to match the infantry's interest in target shooting. Major Generals Oliver O. Howard and Thomas H. Ruger, commanders of the Department of the East during the 1890s, were dissatisfied with the training of gun teams in their region, a shortcoming that Ruger believed was part of a larger problem, the army's urgent need for more artillery. The neglect of the arm was evident at West Point, which during the early 1890s had no artillery horses but only cavalry mounts. Even when these animals were in the best of health, they were too small and light for the rigors of pulling and maneuvering field batteries, and their stamina suffered, as one instructor pointed out, as a result of their being "promiscuously ridden by two hundred inexperienced boys." The Academy's cannon were no better than its horses: the cadets practiced with three-inch rifles and twelve-pound Napoleons, guns of the sort that had defended Cemetery Ridge at Gettysburg a quarter of a century earlier.[17]

It was not just West Point but the entire army that lacked modern artillery. In 1878 General William T. Sherman faulted Congress for not buying the breech-loading rifled guns that he believed the field batteries needed.[18] Ten years later Colonel Robert P. Hughes complained: "Some of the light artillery is still plodding along with the same guns they had at the close of the war of the rebellion, although the Prussians learned from the Austrians at Nachod, nearly a quarter of a century ago, that such guns would not meet modern requirements. [The light artillery] is anxiously waiting for the new steel breech-loading guns." Even the Artillery School had to make do with outdated cannon. One officer protested in 1890: "The guns are all old models and, on account of inherent defects, do not admit of accurate shooting."[19] Obsolete fieldpieces and equipment hindered the school until the eve of the Spanish-American War.[20]

This neglect was particularly unfortunate in that the artillery was likely to assume even greater importance on future battlefields. Lieutenant William E. Birkhimer, a dedicated student of the arm, urged that quick-firing field guns would be more than a match for the magazine shoulder arms of the infantry and that the artillery offered the best answer to the problem of the dominance of the tactical defensive. "Neither the rifle-musket nor machine gun makes any impression on field works," Birkhimer reflected in 1885. "The task of demolishing them devolves on field artillery."[21]

Birkhimer and other gunners found to their frustration that the tactical theory of their arm, as of the others, lagged behind the late nineteenth century's improvements in weapons technology. Birkhimer himself complained in 1880: "It is doubtful if the present system of [artillery] instruction is capable of accomplishing anything of great value." Major L. L. Livingston, commander of the Artillery School, identified the fundamental reason: "The very rapid advances which have been made in artillery material and its application in modern warfare have rendered solid progress in the literature of the subject a matter of extreme difficulty." Colonel John C. Tidball, Livingston's successor, echoed his complaint. "Advancement in the science of artillery has been so rapid of late years," Tidball wrote in 1885, "as to far outstrip the production of text-books, and instruction therein has to be gleaned from many sources."[22]

The Gatling and other machine guns presented an excellent example of technology outrunning tactical theory. Army officers still had not resolved in the 1890s the questions that the introduction of these weapons had raised during the 1860s and 1870s. Should these rapid-firing guns be considered as artillery pieces or regarded as a new type of ordnance? How did they fit into the army's organization? What tactics should govern their use? These questions defied easy answers. Colonel Nelson Henry Davis, a

A Gatling gun and its crew at Fort McKean, Dakota Territory, about 1885. The introduction of these rapid-firing weapons created one of the many tactical debates within the late nineteenth-century American army. (Photograph courtesy of United States Army Military History Institute)

veteran of the Mexican, Civil, and Indian wars, explained the army's fundamental dilemma about these new guns: they were, at one and the same time, both artillery and infantry weapons. "While machine-guns will no doubt . . . form a part of field batteries," he wrote in 1880, "so they will constitute a part of the infantry line and add to its terrible fire." William Birkhimer stated the issue by referring to a sardonic army expression of the day: the "infantry will not have them and artillery does not want them."[23]

Major Edward B. Williston tried as hard as any single officer to resolve these uncertainties about the machine gun's role and organization. During the late 1870s and 1880s Williston commanded Battery F, Second Artillery, which systematically tested a variety of the rapid-firing weapons and tactics for them, the only such experiments conducted before the early 1900s. Convinced that machine guns were uniquely valuable, he advocated in 1886 that they be organized into a separate service. After this proposal was rejected, Williston urged that, at the very least, the infantry-versus-artillery question be resolved by assigning the quick-firing arms to the artillery.[24]

The ambiguity about the organization of the Gatlings and similar arms ran counter to the military principle that each soldier and every weapon should be assigned a clearly defined place within the army, and yet the War Department failed to resolve the issue. Major Williston, as one military scholar later pointed out, could not change by himself the machine gun's status in the army. The official artillery manuals of the 1890s acknowledged the continuing uncertainties about the Gatling and its relatives. The 1896 tactics, repeating language introduced in 1891, bluntly admitted: "The role of the machine gun on the battlefield is not yet determined." "Drills in the use of these weapons have not been obligatory," Brigadier General Wesley Merritt complained in 1894, "because there is no known system of tactics for their use."[25]

Unlike the machine guns, the field guns had a well-defined role and organization, but their tactics were hindered by a different problem, a shortage of current manuals and textbooks. In 1880 the inspector general of the army's eastern posts cited a "want of suitable text-books for the artillery arm." During the winter of 1891 Captain E. L. Zalinski appealed to Tasker Bliss from the Presidio of San Francisco: "I am short of manuals and books for instruction purposes. . . . The men are receptive and would read the books if they had them. Have asked for more but without avail." An officer of the Fourth Artillery complained in 1892 that American gunners lacked guidance and had to turn to foreign sources for advice.[26]

Colonel John C. Tidball, whose long career included battery service during the Civil War and the publication of a manual for heavy guns in 1880, attacked the 1891 artillery tactics while they were still in draft. Tidball criticized the work for assigning field batteries to brigades and divisions, basing his argument on the assumption that an army would nearly always take up a position with openings along its front and that, with rifled cannon replacing smoothbores, it would usually have opportunities for long-range gunnery. A career artillerist, Tidball feared that infantry officers would fritter away these chances. A brigade or division commander standing in reserve or holding a position where the terrain stifled his fieldpieces might fail to detach his cannon to other units where they could contribute to victory. Tidball favored a centralized artillery reserve, commanded by an artillery officer, rather than a decentralized system, in which an army's batteries might be assigned to infantry brigade and division commanders.[27]

Although Tidball presented his case in these terms in October 1890, he privately admitted a few months later to a fellow artilleryman that he was waging what in the twentieth century would be called a "turf battle." The artilleryman confided to another redleg: "There was one point,—one of vital importance to artillery officers,—that I could not with propriety, mention [to the board preparing the 1891 tactics], and that is by assigning batteries to infantry divisions, field officers of artillery are deprived of all command. . . . I want to convey my ideas as fully as possible to artillery officers, and let them see the serpent which is lurking in the grass, poking his head out through the new drill regulations."[28] In a concession to Tidball, the board deleted the provision for assigning batteries to brigades, but retained it for divisions.[29]

While artillery officers debated how their weapons should be organized, they also struggled with the transition from the concept of direct fire to that of indirect fire. Civil War field artillerists had usually delivered direct fire, pointing their pieces by line of sight and firing them at low trajectories against targets at relatively short ranges. After Appomattox, gunnery theory began a transition from direct fire to indirect, characterized by massive, high-angle bombardment from field guns hidden from the enemy. Indirect fire, which envisioned delivering projectiles at high angles over fieldworks, might help end the dominance of the trench, but significant technological improvements were needed, in gun and carriage design, ordnance, explosives, aiming methods, and fire control, to put this new concept into effective practice.[30]

The post–Civil War generation of American artillerists became familiar with the concept of indirect fire. William E. Birkhimer, an enthusiastic student of the history and tactics of his arm, confidently declared in 1885: "Field artillery of the present day is capable of doing excellent work beyond the limit of distinct vision." The official artillery tactics of 1891 and 1896 acknowledged the possibility that "when the target can not be seen on account of smoke which hangs in front of the guns, or on account of fog, rain, or darkness, the pieces may be aimed by means of auxiliary targets." Lieutenant Joseph E. Kuhn, an engineer, translated a European work on indirect fire in 1894. Two years later A. B. Dyer's _Handbook for Light Artillery_ made use of the term, defining it as the fire "from guns with less than service charges, and from howitzers and mortars, at all angles of elevation not exceeding 15°."[31]

American gunners were aware of the concept of indirect fire, but their official tactics did not emphasize it. The 1891 and 1896 manuals envisioned an army's fieldpieces delivering direct fire at ranges extending to three thousand yards, and neither work used the term "indirect fire." American artillerists entered the Spanish War with a gunnery theory that, although in transition to indirect fire, still relied primarily on direct fire, visually controlled by a battery commander. Direct fire by field artillery remained a persistent tenet of the late nineteenth-century army.[32]

The manuals of the 1890s highlighted the importance of visual control. "Correct observation of the effects of fire is necessary," the tactics declared, "in order to make the required corrections [of range], and is indispensable to good shooting."[33] In most cases, this responsibility fell to the battery captains.[34] The manuals counseled against firing at ranges beyond three thousand yards, because although the guns could deliver projectiles that far, the officers could not see and assess the results, "even with good glasses."[35] Once in battle for any length of time, smoke from artillery and infantry fire—which the manuals termed an "inconvenience"[36]—would hinder a captain from observing the effects of his battery's fire. To help solve this problem, the official tactics suggested an idea that was the forerunner of the artillery forward observer team. The 1891 manual recommended stationing an NCO out in front of the guns, beyond one flank, to watch their fire.[37]

In spite of these hints at innovation, the official artillery tactics of the 1890s remained traditional in their reliance on direct fire and on the dashing and daring spirit of the early nineteenth-century light batteries. The 1891 manual boldly advised officers that, in some cases, they should advance their fieldpieces against the enemy's infantry. "Ordinarily, artillery will hold itself beyond the effective infantry fire," the tactics counseled, "but for close support of its own infantry at decisive moments, or before an enemy is disorganized, it should not hesitate to enter this zone and meet the fire of the enemy's infantry at short ranges (800 yards)." The 1896 manual reaffirmed word for word this daring advice.[38]

While these traditional elements endured in the artillery tactics, the future of the cavalry remained a lively subject. In 1884 Lieutenant General Philip H. Sheridan speculated on the nature of combat in the years ahead, envisioning warfare much along the lines of World War I's Western Front. "Armies will then resort to the spade," Sheridan predicted, "the pick, and the shovel; both sides will cover themselves by intrenchments, and any troops daring to make exposed attacks will be annihilated." The hero of Five Forks conceded that these new conditions of warfare would reduce cavalry's tactical role, but he believed that its strategic value would increase. Another famous Union cavalry leader, Wesley Merritt, confidently declared in 1880 that the horsemen had a glorious future ahead of them. Theodore F. Rodenbough, who served nine years with the dragoons and cavalry, reflected in 1889 that the infantry and artillery each had had their "day" and predicted that the third arm would soon enjoy one of its own. "And now we are witnessing the dawn of a cavalry day . . . ," he believed.

"It may be presumed that in the next war the *independence, boldness and versatility* of the cavalry will be marked features."[39]

Despite the dramatic improvements in the firepower of the infantry and artillery, many officers remained enthusiastic about the cavalry's prospects. Philip St. George Cooke, whose career with the mounted service began as an officer in the country's first dragoon regiment, stood among these optimists. Cooke considered tactics "a hobby," published a cavalry manual in 1861, and pursued the subject for decades. He insisted in the 1880s that, despite the improved firepower of infantry and artillery, the mounted arm still could make significant contributions in combat. When a commander gained a chance to pursue a defeated opponent, Cooke urged, cavalry would be of "supreme importance."[40] In 1892 another officer—an artilleryman, not a cavalryman—exceeded Cooke's claim, predicting that because breechloaders fired more rapidly than muzzleloaders, infantry would run out of ammunition on future battlefields, leaving it vulnerable to horsemen. Wesley Merritt flatly asserted in 1880: "There is little doubt that the effect of breech-loading arms is greatly exaggerated." As late as 1897 an officer of the Seventh Cavalry argued that forceful leadership would prove more important than "the many improvements in firearms" and quoted with approval an "old cavalry proverb" that he ascribed to Sir Garnet Wolseley: "Commend your soul to God and charge home."[41]

While some officers maintained faith in the cavalry's future, they continued to debate how the arm should be used. The saber remained at the center of this controversy, continuing to generate emotional arguments. In 1883 Major S. B. M. Young of the Third Cavalry declared that a riding hall was the only place where this edged weapon had any value. "I know of no other use for the sabre in cavalry service," wrote the officer who would become the army's first chief of staff, "except in military honors and teaching the trooper to manage his horse with one hand on drill." Two years earlier, Frank Wheaton had voiced an opinion that was common among veterans of western warfare: "The Sabre in Indian fighting is worse than useless. It embarrasses the Cavalryman and his horse but rarely inflicts as much damage on his enemy."[42] In 1882 General Sheridan flatly pronounced the saber obsolete in the field. While serving as a lieutenant

with the Fifth Cavalry during the late 1880s, the reformer Eben Swift offered a chilling rebuttal to the claim that the saber had the advantage of
being "always loaded." "So is the revolver," Swift asserted and then made
this harrowing suggestion: "If you don't believe it, let any revolver be
pointed at you and see how it feels."[43]

And yet many cavalrymen maintained faith in the saber during the
1880s and 1890s. Some troopers who characterized the edged weapon as
useless against Indians nonetheless believed that it might prove viable
against conventional opponents. "In civilized warfare, I think the sabre is
useful as ever," affirmed Major A. K. Arnold of the Sixth Cavalry in 1880,
"and well sharpened and carried in a wooden scabbard is an effective
weapon, notwithstanding the increases in range of small arms. The sabre
is certainly as effective as of yore, when cavalry is sent in pursuit during
the rout of an army, or in charging a battery, or in charging the enemies
cavalry." In 1892 an artillery officer acknowledged that during Indian
campaigns many mounted commanders had left their sabers at their posts
and taken only firearms into the field. "But this has been due to the peculiar nature of the warfare in which we have been engaged," he believed,
"and not to any distrust of the cavalry officers in the saber as a weapon. . . .
very few of them would consent to the abandonment of the saber." Captain W. A. Rafferty, Arnold's brother officer in the Sixth Cavalry, advocated flexibility: "The nature of the country and warfare can determine
for each commander whether it is better to take his sabres with him into
the field, or leave them stored at his post."[44]

Other officers continued to insist that the saber remained a viable
weapon against any opponent, no matter how armed. A lieutenant of the
Fifth Cavalry contended in 1892 that infantry firing breechloaders rather
than muzzleloaders would be more prone to run out of ammunition and
so would be more likely to leave itself vulnerable to a mounted attack; that
loose formations would be thrown into confusion more easily "than formerly"; and that, with the advent of smokeless powder, defending foot soldiers would have a better view of the horsemen charging them and would
more readily become unnerved. (Moses Harris, another cavalry officer,
commented dryly on this last argument: "This opinion does not inspire

confidence.") Lieutenant Peter Traub of the First Cavalry reaffirmed in 1891 the traditional argument that the weakness of the saber was not the weapon itself but the lack of proper instruction with it. A few years later, a member of the Third Cavalry reiterated another piece of conventional wisdom: "The saber would be much more formidable if ground to a cutting edge."[45]

Guy V. Henry dismissed as "cranks" the soldiers who were still touting the saber during the 1880s,[46] but in fact some highly respected career officers continued to defend the weapon. Lieutenant Colonel James W. Forsyth, a longtime associate of General Sheridan and a careful student of cavalry weapons and tactics, examined some reports from the Military Division of the Missouri and lamented that only one of the eight mounted regiments that were inspected during the early 1880s "carried the sabre in the field, and is the only one in which any systematic attempt has been made to familiarize the enlisted men with this weapon." Another well-regarded career soldier, Colonel John C. Kelton, was prepared to abandon the saber in 1881—but only if the army replaced it with another edged weapon, a "rammer bayonet" that would reach a foot and a half beyond the muzzle of the cavalry carbine. Wesley Merritt, as highly respected as any Civil and Indian wars cavalryman, considered himself a firm believer "in the value of the sabre."[47]

Captain Edward S. Godfrey, another of the army's most admired Indian fighters, and Major George B. Sanford, winner of a brevet at Cedar Creek, helped prepare the 1891 cavalry tactics, which included a manual of the saber and a fencing exercise, intended to develop "the agility, intelligence, and espirit [sic] of the trooper, as well as his adroitness and confidence in the use of the saber." This same work instructed officers that when cavalry met other horsemen, "the object" was "to ride the enemy down, and complete his overthrow with the saber and pistol."[48] When Godfrey reviewed the 1891 tactics for republication a few years later, he was joined by two other highly regarded Seventh Cavalry officers: Lieutenant Colonel Louis H. Carpenter, who had received brevets for the Civil and Indian wars, and Captain Ernest A. Garlington, who had won the Medal of Honor. These cavalrymen endorsed the 1896 tactics, which re-

"Position of Carry Saber, Mounted," a plate from the 1891 cavalry manual. The volume included a drill with this edged weapon and also a fencing exercise with it, intended to develop the trooper's "adroitness and confidence in the use of the saber."

tained an abbreviated version of the 1891 manual of the saber, the fencing exercise, and the advice about riding the enemy down with saber and pistol.[49]

The saber question was part of a larger debate over the merits of mounted and dismounted fighting. Many troopers concluded that improved firearms would force cavalrymen on future battlefields to swing down from their saddles and fight on foot, deploying in loose order and taking whatever cover they could. An officer of the Third Cavalry contended in 1895 that the most important lesson his arm had gleaned from the Civil War was the value of dismounted firepower. Writing about the same time, Captain James Parker of the Fourth Cavalry urged that "the training of the cavalry soldier to fight dismounted is of vast importance." The compilers of the official cavalry tactics of the day agreed with Parker. Recognizing the realities brought by new weapons, the manuals of the 1890s devoted considerable attention to dismounted combat and provided the same instructions to cavalrymen that were given to infantrymen about fighting in extended order on "varied ground." Field experience during the Indian wars may well have contributed to this emphasis on dismounted fighting, just as it promoted the use of loose formations and the improvement of marksmanship. John Gibbon recounted the advice that an "old Indian fighter of the army" had passed on to him: "'When I *once* dismount my men to fight Indians, I *never* remount except to pursue them.'"[50]

The tradition of the mounted charge remained strong, however, throughout the 1880s and 1890s. Colonel Robert P. Hughes complained in 1888 that the "cavalry, as such, does not appear to be found in a very satisfactory condition" and blamed this state of affairs on "a strong tendency towards the condition of mounted infantry in our cavalry." Hughes feared that extensive dismounted fighting during the Civil and Indian wars, and the emphasis on foot target practice during the 1880s, had led "some of our cavalry commanders" to assume "the horse has ceased to be a weapon of the trooper." Cavalryman Moses Harris believed in mounted firing, contending that "the trooper should be able to use his carbine with facility and deliver his fire while moving at a gallop."[51] Joseph C. Breckinridge, the army's inspector general, recommended mounted firepower and mounted shock tactics as well. Breckinridge quoted with approval a German general who had declared, "'The charge should be delivered absolutely en muraille, with no thought of a possible mêleé to follow. The enemy must be made to feel your determination to ride him down, not merely to mingle with him.'" Breckinridge also endorsed the report of an English officer who had observed the charge of a German cavalry unit during an 1889 exercise. "They came on like the fastest rush on a polo ground," marveled the Englishman, "though, of course, the lines were not so well closed up as before. I had to confess, infantryman as I am, that even repeaters would have had a poor chance of stopping the rush."[52]

The 1896 cavalry manual, in the hands of soldiers at the outbreak of the Spanish-American War, advised troopers on mounted tactics against enemy horsemen, infantry, and artillery. Its counsel on cavalry-against-cavalry combat emphasized the traditional horse charge: "In offensive operations, whenever there is opportunity for its employment, the cavalry must be mounted. . . . It is generally expected of cavalry and is its pride to be bold and daring. . . . a sudden and bold charge which surprises the enemy is not only successful, but the loss is small in proportion to its success." Although the manual acknowledged that mounted operations against infantry were best undertaken when the enemy soldiers were out of ammunition, in retreat, or "much shaken by artillery or infantry fire," it continued to provide for the possibility that cavalry would charge against foot troops. The tactics offered troopers the hope that: "Infantry fire will be less steady as the charging force closes upon them." The manual treated mounted attacks on artillery much as it did those on infantry. It conceded that such attempts were most likely to succeed while a battery was limbered, either retreating or taking up a position, but it did not rule out the possibility of such attacks, and it optimistically advised troopers on how to disable captured field guns.[53]

James W. Forsyth stirred a discussion of mounted tactics during the early 1880s, when he urged that cavalrymen devote more attention to mounted target practice. At Forsyth's prodding, cavalry leaders considered this issue and the related question of whether troopers were better armed with the rifle or carbine.[54] Sixteen out of twenty officers of the Sixth Cavalry who were polled on the subject in 1880 and 1881 said that they preferred the carbine. After cavalrymen spent several months discussing mounted target practice and the relative merits of the rifle and carbine, commanding general William T. Sherman resolved the first of these two issues in May 1881, directing that post commanders require their troopers to train in firing from the saddle. The second question was resolved in favor of the carbine, during John Schofield's tenure as commanding general.[55]

Forsyth's study highlighted some of the cavalry's problems, including the need for better horses, improved horsemanship, and more target practice, both mounted and dismounted. One officer in Forsyth's survey made the obvious point that accurate shooting either from the saddle or on foot required practice. The essential question, he went on to say, was whether the troopers would be given the quality of horses and the amount of time needed for marksmanship training. When Eugene A. Carr was asked about mounted firing, the crusty veteran replied skeptically: "There is no probability of our obtaining garrisons of sufficient size and with sufficient immunity from labor, to enable us to properly train our horses and men to fire mounted, particularly as we seem still likely to be condemned to the use of broncos. . . . _Give me my regiment, or as much of it as practicable, together for a year, in garrison and campaign; give me decent horses instead of these miserable brutes, and uphold me in my orders and policy._" "What the cavalryman wants," Captain E. C. Hentig of the Sixth Cavalry believed, "is more practice with his horse and arms, and this can never be secured if the soldier is on fatigue six days out of every seven."[56]

The cavalry was not alone in its problems. Artillery officers tried, with varying degrees of success, to overcome the general neglect of their arm, to improve their tactical literature, to grapple with questions of organization, and to complete the slow transition from direct to indirect fire. Infantry theorists struggled with the daunting challenge of how to cross the "deadly ground" at the same time that conservative ideas persisted about the bayonet and musketry. As these issues grew in number and complexity, thoughtful soldiers recognized that the army needed not merely drill books but comprehensive tactical manuals.

6

*Great Changes Now
and to Come*
The Leavenworth Board
Manuals, 1891

The three "assimilated" drill books that the army had
adopted in 1874 served as its official tactics throughout
the 1880s. Between 1875 and 1890, D. Appleton and Company offered no fewer than nine editions of the trilogy's
infantry manual, Emory Upton's *Infantry Tactics, Double
and Single Rank*, each identical to the 1874 original. During the same fifteen years Appleton also reissued the assimilated cavalry volume at least three times and its artillery companion seven. In 1887 Colonel George Gibson of
the Fifth Infantry compiled a *Battalion Drill and Battalion
Skirmish Drill*, a small-unit infantry tactics derived from
the assimilated system.[1]

The 1874 manuals were supplemented by works like
Gibson's and by War Department directives answering
specific questions about them, which were published in
the *Army and Navy Journal* and *Army and Navy Register*.
These interpretations in professional publications helped
the assimilated volumes remain the army's official tactics
for seventeen years, but the number of questions about
the 1874 system, and of recommendations for changing it,
grew embarrassingly large.[2] The *Army and Navy Journal*

commented in 1888 that the proposals to amend the assimilated tactics would "fill a Saratoga trunk." The War Department issued so many clarifications of the 1874 volumes that one officer complained in 1890 that this mass of interpretations had become "the bane of military instruction in tactics during the last five and twenty years."[3]

While critics filled the columns of the *Army and Navy Journal* and *Register* with evidence of the shortcomings of the 1874 manuals, some officers concluded that the time had come for the army to move beyond the drill books of the sort that William J. Hardee and Emory Upton had compiled, to a more broadly conceived system of tactics. Upton himself stood among their number. "You are well aware that thus far in our history," he wrote to General William T. Sherman in 1880, "Tactics, in all arms of the service[,] have been simply a collection of rules for passing from one formation to another. How to fight has been left to actual experience in war." During the 1870s Upton had observed armies around the world and had been impressed particularly by the small-column maneuvers that he saw performed by Indian, Italian, Russian, and Austrian units. Upton had experimented with this formation during the Civil War and had later become skeptical of its utility until these foreign studies revived his confidence in it. He began work on a new tactics that featured small columns, with the companies of an infantry battalion moving into battle in double columns. This effort was left unfinished by Upton's suicide in March 1881, but before his death he indicated to Sherman that the work he envisioned would have been more than a drill book. "After finishing the tactics proper," Upton reflected, "I think it would be desirable to have a chapter on 'applied tactics' i.e. how to get into line of battle with a company[,] platoon[,] regiment[,] brigade &c, how to turn a flank, how to meet flank attacks, to occupy ground, make or procure shelter & etc."[4]

Other soldiers agreed that the army needed a more comprehensive tactics. While preparing an 1884 inspection report, Lieutenant Colonel N. B. Sweitzer of the Eighth Cavalry drew a distinction between "maneuver tactics," the close-order drill movements that would bring a unit to the battlefield, and "fighting tactics," the loose-formation operations that it would carry out in combat. Two years earlier, another inspector had urged that less attention be given to linear formations and more to the ability of small units to maneuver in double columns. In 1887 Brigadier General George Crook declared that the army's infantry-attack formations would not stand the test of war and new ones were urgently needed.[5]

This interest in tactics came during an era when the army was gaining greater professionalism, a trend evidenced by the growth of its schools. The Artillery School at Fort Monroe, founded in the 1820s and active until 1860, was revived in 1868. During the early 1880s the Willett's Point, New York, station was transformed into the Engineer School of Application. General Sherman established a School of Application for Infantry and Cavalry at Fort Leavenworth in 1881, and ten years later General Schofield ordered every post that was garrisoned by line troops to establish an officers' lyceum.[6] Service associations and the journals that accompanied them also furthered the army's professionalism. The *Journal of the Military Service Institution of the United States* began publication in 1879 and the *Journal of the United States Cavalry Association* in 1888. "From these centers," Matthew F. Steele, whose own career embodied the army's new professionalism, wrote in 1895, "a wave of professional culture has swung out in widening circles, until to-day it reaches every little garrison and well nigh every individual in the military service." Officers read a number of other military periodicals, including the *Army and Navy Journal*, established in 1864, the *Army and Navy Register* and *United Service*, in 1879, and the *Journal of the United States Artillery*, 1892. These publications featured articles on war gaming, map exercises, field maneuvers, and tactics, and they gave American soldiers the chance to read the work of British soldiers and translations of European writers.[7]

The appearance of foreign articles in these journals did not mean that Americans invariably wanted to copy European ideas about tactics and other military subjects. The truth was quite the opposite: United States Army officers remained as strongly nationalistic during the 1880s and 1890s as they had been during the first fifteen years after the Civil War. "Almost anything is better than European tactics," Captain Charles G. Ayres declared in 1894. When Lieutenant Colonel Hamilton S. Hawkins drafted an infantry manual and lobbied for its adoption in 1890, he claimed for his own work one of the appeals made for Upton's, that it was "an *American* system of infantry drill." A year later, an officer arguing that the

The Class of 1891 at the Infantry and Cavalry School, Fort Leavenworth, Kansas. The army gained greater professionalism during the 1880s and 1890s, a period of growth for its schools and journals. (Photograph courtesy of United States Army Military Institute)

army should develop mounted infantry units believed he had made a deci-sive point when he reminded his audience that cavalrymen had often fought dismounted during the Civil War. "*Be American*," this soldier ad-vised, "Follow the teachings and experience of our own great leaders." In 1892 Lieutenant Charles D. Parkhurst expressed regret that Civil War ar-tillery veterans had not published their ideas about the tactical lessons of that conflict, forcing the next generation of United States Army officers to turn to foreign sources. Two years later one particularly distraught Ameri-can soldier fired off a letter to the *Army and Navy Register*, listing six British textbooks that "patriotic Americans have to study" at their military schools. "It is lucky . . . ," he noted sarcastically, "that so far no Englishman has developed brains enough to get up a book on 'ballistics,'" or else United States officers would be studying an English text on that subject as well. "It is full time," this unhappy correspondent pronounced, "to call a halt in this direction. Give us our American text books."[8]

Many United States officers wanted their army to have a system of tac-tics of its own, independent of European models, and others saw that new ideas were demanded by improvements in weapons. As repeating breechloaders, with their more rapid fire, replaced the single-shot muzzleloaders of the Civil War and made attacking even more dangerous than it had been during that conflict, some theorists realized that adequate tactics would require more than having battalions practice drilling in lin-ear formations. "Breech-loading guns have introduced great changes," the veteran artillery officer Henry W. Closson observed in 1891, "and doubtless there are others to follow in the direction of rapid fire and metallic case ammunition." Drawing on a British source, an 1880 *Army and Navy Journal* article claimed that infantry had become "beyond all question the decisive arm" and its fire "the great arbiter" of battle. One analyst raised the specter that "while foreign nations are preparing to use their infantry fire effec-tively at very long ranges, we in America are behind the age."[9] N. B. Sweitzer believed that his country's soldiers needed tactics that would al-low them to carry their breechloaders easily and to use them flexibly, whether attacking or defending.[10]

Still another impetus for change came from a discontent with Upton's "fours," the groups of four men that formed the basic units of his system.

This element of Upton's tactics had sparked a controversy when it was introduced in 1867, and criticism of it had continued into the 1880s. Its structure was cumbersome: after a company divided into groups of four, there might be an odd file of two soldiers (a "half four"), or an extra man, or both, left over.[11] Several officers pointed to an even more serious drawback, the likelihood that the system, in which each soldier was assigned a number— one through four—within his group, would break down under the stress of combat. As men were killed and wounded, their surviving comrades were to renumber and maintain the sets of fours, leading one sarcastic critic to suggest: "Keno-callers would make good captains for this sort of business."[12] Lieutenant Colonel Henry M. Lazelle, who had experienced the confusion of combat during the Civil War, discerned the same weakness in Upton's blocks of fours: "Loss of men [in battle] at once annihilates . . . the possibility of reformation or continuance of any formation [based] on that system. It is only a drill and dress parade system." In 1888 Major John A. Kress, a long-tenured veteran, was even harsher than Lazelle on Upton's work, which he characterized as "a source of confusion at drill," which "falls to pieces completely in action when the frequent changing of numbers and counting off becomes ridiculous in the extreme.—It is an absurdity."[13]

Some officers believed, too, that the time had come to discard Upton's assimilated system, which applied to all of the arms of service, and to refine the tactics of the infantry, cavalry, and artillery as distinct entities. An 1885 correspondent to the *Army and Navy Journal* flatly pronounced the assimilated system "a fraud. . . . We want infantry tactics for infantry and a long rest after [the] incessant wrangling [over tactics]." William E. Birkhimer of the Third Artillery, a devoted student of the history and tactics of his arm of the service, contended in 1884 that assimilation had been detrimental to the artillery. Four years later, William H. Powell of the Fourth Infantry advised abandoning the system.[14]

The army's senior officers agreed with their subordinates. In 1886 inspector general Absalom Baird urged that a revision of the army's tactics was "much needed," with "the instruction for each arm of service . . . treated according to the requirements of that arm by itself, without regard to a forced conformity to the school of another arm." A year later Lieutenant General Philip H. Sheridan, the commanding general, declared that the army's fourteen years of experience with the assimilated tactics had been disappointing. He judged that the 1874 system had proven unnecessary, "tending to unduly limit the . . . especial individuality which each arm should retain if it is to be employed to the best advantage." Sheridan's successor, Major General John M. Schofield, in 1890 expressed his hope that the time had arrived when "the instruction of each arm of the service may be more closely confined to that which will make the troops most efficient in their own special service."[15]

During the 1880s several officers offered alternatives to the Upton system, none of which gained the sanction of the War Department. In 1882 William H. Morris, a Civil War brigade commander with a long interest in tactics, offered a *Tactics for Infantry Armed with Breech-Loading or Magazine Rifles*, which provided for infantry to maneuver in two ranks and to fight in one. Morris revised this work six years later and lobbied commanding general John Schofield to endorse it, but the author failed in his campaign to make his volume the army's authorized tactics.[16] Nor did the War Department adopt either "A Manual of Drill and Tactics," offered in 1883 by Lieutenant G. N. Whistler of the Fifth Artillery, or Captain John H. Patterson's 1884 *Infantry Tactics*, which emphasized loose order rather than "perfection of alignment on the firing line" and prompted company officers to "encourage the individuality and self-reliance of their men."[17] Another captain, Robert P. Hughes, Brigadier General Alfred H. Terry's brother-in-law and his aide during the Little Big Horn campaign,[18] prepared an infantry tactics that Terry recommended to commanding general Philip H. Sheridan in January 1884. Hughes addressed the problem of increased firepower in the same way that Hardee had tried to deal with it in the 1850s, by speeding up the movements of the foot soldiers. Hardee's solution had proven inadequate during the Civil War, and one or more senior army officials apparently feared that Hughes's proposal would also fail the test of combat. The work was supported by Terry, Sheridan, and Wesley Merritt, but it did not gain the War Department's sanction.[19] The same proved true for many other proposals prepared during the late 1880s by Hamilton S. Hawkins,[20] William R. Livermore,[21] Frank H. Edmunds,[22] Edward J. McClernand,[23] and many other officers.[24]

The army did not begin to leave the Upton era until January 1888, when Secretary of War William C. Endicott authorized a board of officers to review the tactics of each of the arms of the service. Lieutenant Colonel John C. Bates, who had been a company officer during the Civil War,[25] chaired this panel and First Lieutenant John T. French, Jr., of the Fourth Artillery served as its recorder. Six other officers, two from each of the three arms of the service, completed the board's membership. The most famous among the group was Captain Edward S. Godfrey, a highly respected veteran of the Little Big Horn and other Indian campaigns.[26]

When the board began meeting in Washington, D.C., in February 1888, it faced what the *Army and Navy Journal* called a "formidable task." The panel was soon swamped with proposals from theorists who believed that their ideas deserved the War Department's endorsement. By early March the board had received five drafts that were revisions of Upton's work and three entirely new systems, and submissions continued to arrive nearly every day. One officer alerted the board that he would be submitting a proposal, another sent in a general prospectus, and at least one civilian offered his ideas.[27]

The panel continued to meet in Washington until late March 1889, when it moved to Fort Leavenworth, Kansas, to be close to the School of Application for Infantry and Cavalry, which had been established there in 1881. Leavenworth's officers found no room on the post for the board and it moved into offices and quarters nearby in the town.[28] Here the group finished its work in December 1890 and the next month sent final drafts of tactical manuals for infantry, cavalry, and artillery to the adjutant general.[29]

At an earlier, simpler time in the army's history, the commanding general himself would have evaluated the proposals of a board like this one. By the late nineteenth century, staff procedures had become more sophisticated: before these 1891 submissions went to General Schofield, they were closely reviewed by two brigadier generals and a colonel. Each of the three reviewers assessed for General Schofield the volume that would be used by their particular arm of the service. Brigadier General Wesley Merritt replied first, endorsing the cavalry tactics on March 17, 1891. Colonel Henry W. Closson completed his review of the artillery volume later the same month, and Brigadier General Thomas H. Ruger commented on the infantry manual in July. With the reports of these three Civil War veterans before him, General Schofield approved the work of the Leavenworth board that autumn. His endorsement was followed quickly by those of the secretary of war[30]—the reform-minded Redfield Proctor[31] then held the office—and the president.[32]

The review of the Leavenworth manuals was more thorough than that given earlier ones because the army had become increasingly bureaucratic during the late nineteenth century. Another, more subtle example of this change can be discerned in the way that soldiers referred to the new tactics. Although they were reviewed by groups of officers, earlier manuals were invariably closely associated with the individual who had prepared them. Soldiers referred to these works as Scott's tactics, Hardee's tactics, or Upton's tactics. Lieutenant John T. French was the soldier most closely associated with the 1891 infantry manual. He served as recorder of its board, published a portion of the work in the *Army and Navy Register*, and compiled a series of interpretations of the volume for the *Army and Navy Journal*.[33] Yet American soldiers did not call the 1891 manual "French's Tactics," or "Bates's Tactics," for the chairman of the Leavenworth board. They referred to it in a less personal and more institutional way, as "the new drill regulations" or "the infantry drill regulations."[34]

The new manuals went promptly into the hands of army officers. General Schofield accepted the suggestion of a Union cavalry hero, Wesley Merritt, that the *Journal of the Cavalry Association* publish a preliminary draft of the tactics for the mounted arm. This new periodical serialized the work of the Leavenworth board, offering horse soldiers an early look at the tactics while they were in press with the Government Printing Office.[35] The *Army and Navy Register* did the same with the infantry manual, running an introductory version over the course of a dozen issues through the autumn of 1890.[36] The Government Printing Office published the official first edition, titled *Infantry Drill Regulations, United States Army*, before the close of 1891. One enterprising retired officer, Hugh T. Reed, bought the electrotype plates that the Government Printing Office had used to illustrate the volume, abridged the text, and paid to print his own version, getting it into circulation also before the end of 1891. The New York–based D. Appleton and Company published the Leavenworth infantry manual at least four times before the Spanish-American War: in 1891 through 1893 and in 1895. For Appleton's 1893 edition, the editors of the *Army and Navy Journal* prepared an "interpretations" section, answering questions that had been raised as officers gained experience with the new tactics.[37] Although the artillery and cavalry volumes did not enjoy as many press runs as the infantry manual, they were made available quickly, appearing in print before the end of 1891.[38]

The Leavenworth manuals, which endured to serve as the army's authorized tactics during the Spanish-American War, included significant improvements over the old Upton and assimilated systems and represented a landmark effort. Until the adoption of the 1891 volumes, the United States Army never had a "tactics," properly defined, a manual that would instruct officers how to maneuver and engage their troops to gain advantages in battle. "What we call 'Tactics,'" wrote one artilleryman in 1887, "no other nation in the world dignifies with that name." The volumes that the War Department had approved up through Upton's works were more accurately called "drill books" rather than "tactics." They defined the formations and movements for the various types and sizes of units, but they did not advise officers—as Upton himself had put it—"How to fight." Tactical theorists had offered their opinions on this broader subject in any number of private publications, but no official work before the Leavenworth tactics of 1891 gave American soldiers such advice.[39]

This separation of "drill" from "tactics" was related to the distinction, made by N. B. Sweitzer in 1884, between "maneuver tactics" and "battle tactics." "Maneuver tactics" referred to the close-order drill movements that would bring a unit to the battlefield, while "battle tactics" were the loose-order operations that it would conduct in combat. American soldiers sharpened this distinction during the 1880s. From ancient times, units had always been able to march in one formation and to fight from another, but as Lieutenant Lyman W. V. Kennon observed in 1888, improvements in weapons now made it more important than ever that troops moving to the battlefield in close formations be able to deploy for combat in loose order. "The distinction between manoeuver tactics and fighting tactics," Kennon claimed, "is more marked than it ever has been before." "We shall have two kinds of drill in future," another infantry officer predicted in 1889,

Company B of the Seventh Infantry at Fort Logan, Colorado, during the late 1890s. This photograph shows infantrymen marching in a column of fours, training with the tactics the army had adopted in 1891. (Photograph courtesy of United States Army Military History Institute)

"manoeuvre tactics and battle tactics; the former consisting mostly of a few simple movements, in more or less closed formations; the latter of open order movements, which will be used under fire, or in the immediate vicinity of the enemy." This prognostication was realized in the Leavenworth 1891 infantry manual, which consisted of two parts: a "Close Order" that used a variation of Upton's groups of fours as the formation for bringing troops to the battlefield and an "Extended Order" that deployed them loosely apart for combat.[40]

The "Close Order" section of the new manual developed its own version of Upton's fours and used this formation for moving troops into battle. In this part of the infantry book, the term "four" actually meant "four files." A file was a soldier and his comrade posted directly behind him, so that the "fours" of the close-order tactics consisted in fact of eight men, a corporal and seven privates. The "Extended Order" section of the new manual used the less confusing term "squad" to refer to these eight men, after they had deployed for battle.[41]

While making use of the "fours," the 1891 "Close Order" tried to resolve some of the formation's problems: what to do when there were missing men, or extra men, or an odd number of fours, or how to compensate for lost soldiers. The Leavenworth tactics allowed a unit to march in fours and maneuver in two lines, and it addressed the "missing man" dilemma by providing that if a soldier were missing from the front rank, he would be replaced by the man from the line behind him. Two solutions were offered for the problem of extra men: the more practical answer used them as file closers, soldiers who followed in the rear of an attack and tried to prevent straggling or deserting. (The alternative proposal required replenishing a set of fours by a formula so complicated that it was no better than Upton's system.)[42] If a company had an odd number of fours, the "extra" set of fours was assigned to the right half of its front. As for the most daunting of these questions, what to do when a set of fours became reduced, the 1891 tactics offered a more realistic solution than Upton's renumbering under fire: it used the fours only for marching and maneuvering, not fighting. There remained the problem that a company might lose soldiers to straggling or distant artillery fire and that the fours could break up even before they reached the battlefield. Acknowledging this unhappy possibility, the

Leavenworth volume offered a short-term solution of having the men "fall in without regard to fours, but in their relative order, closing to the right, so as to leave no blank files." Later, presumably when a lull in the engagement allowed it, the company's first sergeant would call the roll and supervise the renumbering of its fours.[43]

While the "Close Order" used the groups of fours for marching and maneuvering, the "Extended Order" featured what was perhaps the most important single innovation in the new volumes, basing the battle tactics on loose-ordered squads of eight men each. Hardee's and Casey's manuals, which still contained much of Scott's musket tactics, had deployed Civil War soldiers for battle in close-ordered, two-line formations. Four years of bloody fighting demonstrated the range and accuracy of the rifle, and after the war Upton offered the army a single-rank tactics, on the premise that the firepower of one line of rifles could be as destructive as that of two lines of muskets. As breech-loading repeaters presented an even greater challenge than the Civil War's muzzle-loading rifled muskets, the 1891 tactics took another step beyond Upton's single-rank formation. Its "Extended Order" deployed soldiers into loose formations, spreading them out to minimize the targets they would present to the destructive firepower of breech-loading repeaters, Gatling guns, and improved artillery.[44]

The new manual provided that the "squad is the basis of extended order. . . . This instruction, on account of its importance, will be given as soon as the recruits have had a few drills in close order." The *Army and Navy Register* explained this dramatic innovation to its readers in September 1891: "With these tactics the line of battle disappears. . . . The fighting line will consist of a series of squads as skirmishers, the normal number of men in a squad being eight." The *Register* had recommended this formation a year earlier, on the grounds that it represented the most recent tactical thinking and had been endorsed by European authorities.[45]

The Leavenworth manual called for infantrymen to practice the new extended-order operations first on a drill field, then "on varied ground," with the soldiers "making use of the accidents of the surface for cover, etc., and observing the conditions of battle." A company officer was expected to take his men into the field and to explain the relationship between terrain and tactics, analyzing "the form of the ground and the different military purposes to which its features are adaptable, using and explaining the military terms that apply; he will require the men to point out the leading features of the country in sight or near their position, with all that concerns the streams, roads, woods, inhabited places, etc." This training shifted from one area to another around a post, "to accustom the men to new situations."[46]

The 1891 manual suggested that an officer point out the direction of an imaginary enemy advance and have his soldiers practice choosing their own positions, picking out the safest cover, keeping a view of the opponent's line of approach, and staying within sight of their comrades. After the men had learned to find and use shelter, they could practice advancing in their loose formations, making short advances from one covered point to the next. The Leavenworth board tactics recommended an exercise in which one soldier sallied six hundred or so yards in front of his comrades, taking a position to represent the enemy's line. Then the infantrymen individually practiced advancing in rushes against their mock opponent. "In order to keep out of sight of the enemy," the manual advised, "the recruit must make the best use of cover, but must not deviate too much from his direction; he must stoop and even creep or crawl, but, if possible, never lose sight of the enemy; open ground exposed to the fire of the enemy should be crossed at a run, by rushes of about thirty yards, then taking the lying position and raising the head in order to see the enemy." Private Charles Johnson Post recalled that his regiment, the Seventy-First New York Volunteers, took this sort of training after the outbreak of the Spanish-American War. Although Post disparaged nearly all drills, he acknowledged that his unit practiced one "substantial innovation," the advance by rushes.[47]

The 1891 tactics called for practicing defense, as well as offense, in loose order. "A well-instructed soldier or non-commissioned officer is then placed in the position of the enemy," the manual provided, "and required to advance upon the skirmishers; the later will carefully observe his movements and aim at him when he exposes himself, adjusting the sight to agree with the distance." Cavalrymen, who often had to fight dismounted, were expected to develop the same skills.[48]

The 1891 infantry volume distinguished between close-ordered formations for maneuver and loose-ordered ones for battle, and it was also the first tactical manual, rather than drill book, in the army's history. It advised commanders how to fight, addressing the distinctive problems of attacking and defending. These two combat roles had emerged, perhaps, when one primitive human first assumed the tactical offensive by picking up a stone to throw at another, who took the defensive by hiding behind a boulder and hurling rocks back at his antagonist. From ancient times on, attackers had held the advantages of the initiative and the element of surprise, while defenders, free of the burden to act, enjoyed time to protect themselves and respond to their enemy's moves. For centuries the relationship between offense and defense shifted, as weapons improved and tactics changed.[49] In American military history, the advantage had swung sharply to the tactical defensive during the Civil War, when rifled shoulder arms and increasingly sophisticated field entrenchments dominated the battlefields of both theaters.[50]

The Leavenworth board manuals, published nearly a generation after the clash between the North and South, represented the first official works to discuss the distinctive problems of offensive and defensive fighting. The 1891 artillery and cavalry volumes discussed how each arm would perform these two roles. The infantry work contained separate sections dealing with offense and defense, at the company and battalion level. This manual advised a captain how to maneuver his company within its battalion during an offensive, in particular how to relieve other companies under fire and also how to handle his unit if it were attacking alone. Similar guidance was given about how a company should fight on the defensive, either as part of its battalion or on its own. The company tactics also counseled infantry officers about attacking and defending against cavalry and artillery.[51]

Battalion tactics received much the same treatment. The manual advised commanders how to handle their unit on the offensive or the defensive, acting either in concert with other battalions or alone. A section on "Dispositions of a Battalion" covered topics including how a battalion might operate on a flank—to secure it, to turn the enemy's flank, or to divert the opponent's attention while the main attack was made elsewhere. The same portion of the tactics advised a battalion commander about serving as an

"The Whistle," an advertisement in the *Army and Navy Register*, January 13, 1894. The Pettibone Company's artist depicted the squad formation that the 1891 manual had introduced as the basic infantry unit. Seven soldiers advance in loose order, followed by their sword-bearing NCO.

advance guard, attacking and defending against cavalry and artillery and operating at night.[52]

Upton's, Hardee's, and other drill books before 1891 had described how units were to form and move, but as to how to conduct troops in battle, they had offered either minimal, commonsensical guidance or none at all. The new manual addressed this need, offering commanders advice about specific problems of both offense and defense. It suggested, for example, that a captain whose company was moving into an attack with its battalion should form his men in line and lead them forward until they came under enemy artillery fire and should then send ahead a noncommissioned officer with a few scouts. An advancing company would hold its fire as long as possible, relying first on its sharpshooters to develop the enemy's position and volleying only against defenders foolish enough to come out into the open. When a captain saw that his unit could no longer sustain its attack without firing, he would call out the number of volleys to be delivered at each halt. Battalion commanders were advised how to deploy to defend artillery: "The companies are posted in front of the intervals between groups of batteries and on the flanks, so as not to hinder the fire of the artillery; they are held ready to meet the attack."[53]

To improve the chances of sustaining an attack in the face of heavy fire, the Leavenworth board tried to give field officers more direct control over their men than they had exercised during the Civil War. The 1891 tactics introduced three new organizations, below the level of the company: the platoon, section, and squad. A company was made up of two platoons, each commanded by a lieutenant. A platoon consisted of two sections, which were administrative rather than operational units.[54] The most important of these three organizations was the squad, a combination of two sets of fours, eight men who became the fundamental unit in the new, extended-order tactics. These seven privates and a corporal, loosely deploying and advancing in short rushes from one point of cover to the next, were the basis of the Leavenworth board's small-unit offensive tactics.[55]

The 1891 tactics also carried the advantage of being compatible with the proposal that regiments be divided into three battalions, an organizational scheme that addressed the problem of maintaining the offensive against more accurate and rapid infantry and artillery fire. Throughout the nineteenth century, the army had defined a battalion as a collection of two or more companies. As many as all ten companies of a regiment might form a battalion, as had been the common practice of Civil War volunteer units, making a "battalion" and "regiment" synonomous.[56] After the War Between the States, many military leaders urged that the army should move from a single-battalion regiment composed of ten companies to a three-battalion regiment, with four companies per battalion, twelve companies per regiment.[57]

Secretary of War Daniel S. Lamont in 1894 carefully explained the need for this reform, detailing how changes in firepower demanded both new tactics, deploying soldiers in loose order across extended fronts, and also a new regimental organization. He reviewed the conditions during Civil War combats: "Formerly, and down to a recent date, the colonel could see and direct the movements of all the men of his regiment who marched and fought in double rank with touch of elbows. Under such conditions a regiment of 1,000 men occupied a front on the battle line no greater than would now be covered by a small battalion of one-third that number." Lamont then contrasted the old and new circumstances on the battlefield: "A few years ago small-arms fire was ineffective at distances greater than 600 or 800 yards, while now it will be deadly at ranges of 2,000 yards, or at even greater distances. In modern warfare the men will act in small groups or singly, and the advance will be made in successive lines in open order." The secretary built a compelling case that a battalion of four companies represented the largest body of troops that a single officer could control in battle.[58]

Upton had predicated every edition of his infantry tactics on the ten-company, single-battalion regiment, but late in his life he favored both the twelve-company, three-battalion regiment and also the company column, which was compatible with it.[59] A long series of commanding generals—Grant, Sherman, Sheridan,[60] Schofield,[61] and Miles[62]—endorsed the three-battalion regiment, as did secretaries of war Proctor,[63] S. B. Elkins,[64] and Lamont.[65] The chief obstacle was money: in 1894 the army consisted of twenty-five regiments and Congress hesitated to fund the additional two companies per regiment that the three-battalion proposal required.[66] Concerned about these costs, the national legislators resisted this reform until after the outbreak of the Spanish-American War. By then the

Leavenworth tactics, harmonious with the three-battalion organization, had been in place for seven years.[67]

The 1891 manual also proved compatible with the army's first smoke-less-powder, magazine rifle, the Danish bolt-action, .30-caliber Krag-Jorgensen. The army adopted a repeating rifle only after years of cautious study. "If we have been slow in this matter," Secretary of War Proctor explained, "we have at least been saved the great expense incurred by most foreign nations in adopting an unsatisfactory magazine arm prematurely."[68] The army selected the Krag-Jorgensen in 1893 and armed seven regiments with it by November of the next year. In 1897, on the eve of the Spanish-American War, the War Department supplemented the 1891 tactics with a manual of arms for the Krag.[69]

Many officers praised the new infantry volume. "In our judgment," the *Army and Navy Register* declared in its first review of the work, "the board has done good work and has been guided by sound and safe principles." The *Register*'s reviewer deduced that the Leavenworth panel had arrived at its "Close Order" by revising Upton and prepared its "Extended Order" by drawing on French, and, to a lesser extent, Belgian and German, sources. He touted in particular the use of squads in open order as the basis of the battle tactics. A later review in the same journal complimented the Leavenworth board "that in the matter of tactics it has effectually prepared us for war." Lyman W. V. Kennon, an aide to George Crook who shared the general's interest in tactics, joined the *Register* in praising the new manuals.[70]

Much of this enthusiasm for the Leavenworth volumes was well deserved. They proved flexible, and they replaced the assimilated system with an improved tactics for each individual arm. William H. Carter, in 1894 a captain with the Sixth Cavalry and later a general officer, touted the cavalry manual as the most clearly written drill book to date. Another captain, Frederick K. Ward of the First Cavalry, commended the emphasis the new tactics gave to extended order, noting that in the past troops had mastered the formation under fire rather than in peacetime. Ward considered the Leavenworth board's cavalry volume, like that on infantry, to be the army's first true tactics. "All previous systems of tactics used in our service that I have any knowledge of," he reflected in 1893, "were limited almost entirely to drill movements. Though they were called tactics they had very little of tactics in them." The captain commented on the irony of the 1891 manual's title, *Cavalry Drill Regulations, United States Army*, and his remarks applied equally well to the infantry and artillery volumes. "Our new book, called drill regulations," he observed, "might without impropriety be called tactics. For the first time a manual is adopted which carries the instruction far enough to really prepare the soldier for the battle field."[71]

Like the 1891 cavalry manual, the artillery tactics abandoned the assimilated system and instead treated the field guns as a separate arm of the service. The Leavenworth Board made only a few changes in the drill commands of the 1874 artillery work while adding sections on the 3.6-inch field gun and the machine gun. The most significant difference between the 1891 and 1874 artillery volumes was the addition of forty pages on "Artillery in the Field," a collection of advice that made the 1891 artillery work, like its infantry and cavalry companions, a tactical manual rather than a drill book. This section advised artillerymen on subjects such as zones of fire, offensive and defensive fighting, combat against infantry, fire control, and machine guns.[72]

In addition to these important changes, the Leavenworth board tactics featured other improvements over their predecessors. The 1891 volumes were introduced by three and a half pages of definitions, including this practical entry on "tactics": "The art of handling troops in the presence of the enemy, *i.e.*, applying on the battle field the movements learned at drill." The Leavenworth board manuals were also the first in the army's history to counsel commanders about maintaining hygiene in the field, recommending that soldiers use tree branches, leaves, or straw to raise their beds off the ground and that they fight diseases with cleanliness and proper cooking. The 1891 tactics also addressed directly the important but unpleasant subject of latrines, which earlier works had treated with Victorian silence. "On arriving in camp," the manual told officers, "if orders are not communicated for resuming the march the following morning, sinks should at once be dug. The sinks should be concealed by bushes or tents, and should be covered daily with fresh earth."[73]

The 1891 manuals, which nudged the army closer to the twentieth century, were supplemented by a number of other works on strategy and tactics that were available to American officers during the 1880s and 1890s. The textbooks used at the Artillery School in 1886 and 1887 included Hamley's *Operations of War*, Jomini's *Summary of the Art of War*, and a pamphlet by the British theorist Lumley Graham, "Tactics of Infantry in Battle." At the Infantry and Cavalry School, students in the Department of Infantry in 1893 read a British work, Lieutenant C. B. Mayne's *Infantry Fire Tactics*, as well as the Leavenworth Board infantry manual,[74] while officers in the Department of Military Art studied Hamley. The next year the scholarly Captain Arthur L. Wagner took charge of the Department of Military Art and taught from his own works, *The Campaign of Königgrätz* (1889), *The Service of Security and Information* (1893), and *Organization and Tactics* (1895), stimulating an interest in tactics among the Leavenworth students.[75]

These books, and the 1891 tactical manuals, appeared at a time when thoughtful American officers perceived that improved weapons had changed the realities of combat. The Leavenworth volumes represented a milestone in the army's thinking about tactics. Their predecessors, from colonial times through Upton's works, were no more than drill books. The 1891 volumes were the United States Army's first true tactical manuals, providing officers with advice on how to fight, on both offense and defense. It remained to be seen how well these new tactics would fare against the complaints of critics and, later, the tests of battle.

7

No Final Tactics
Questions without
Answers, 1880–1898

Many officers criticized the Leavenworth board manuals even while they were still in draft. The first reviewer of the infantry volume, Brigadier General Thomas H. Ruger, endorsed the work but qualified his recommendation with a number of reservations, particularly about the "Extended Order." General Ruger cautiously suggested that the army issue a provisional edition of the new tactics and allow officers to make suggestions about the work before its final acceptance. Reversing Ruger's criticisms, another soldier welcomed the innovative extended-order formations but faulted the close-order ones. Comparing the two-part volume to "an ill made pair of boots, one of which fits and the other does not," this critic endorsed the extended-order portion but called the close-order section "bad—so bad that it ought to go into the waste-basket." Colonel Henry Lazelle was also among those who sniped at the work of the board, even before its publication. He questioned the credentials of the panel members and criticized their methodology, as well as their results.[1]

The criticism increased after the War Department

adopted the Leavenworth tactics. Some officers feared that the new manuals were too long and complicated to be mastered by the large volunteer forces that the country would have to train quickly at the outbreak of war.[2] Colonel Thomas M. Anderson of the Fourteenth Infantry acknowledged that the Leavenworth board had been "most painstaking and industrious" but published an article that was largely critical of their results. "None of the members of the Board were tactical experts," asserted Colonel Henry M. Lazelle, who went on to fault them for introducing changes, "apparently only for the sake of change." "In many instances," another officer complained a year after the tactics had been in print, "clearness of expression has been sacrificed to brevity." Inspector General Joseph C. Breckinridge reported in 1893 that most officers agreed with General Ruger's initial criticism of the tactics that the close-order drill was satisfactory, but the extended order was not. Breckinridge thought the prevailing view was that the new loose-order tactics required too many leaders and gave them too much authority and that under the stress of combat the system would break down in confusion.[3] The inspector general also concluded that most officers agreed with another of Ruger's original criticisms, that abandoning the single-rank tactics had been a precipitous mistake.[4] In addition to these broad objections, as the years passed and soldiers gained more experience with the details of the new manual, some of them quibbled with its specific formations and movements.[5]

The cavalry and artillery works drew criticism as well. Lieutenant D. L. Brainard complained in 1893 that the new cavalry manual was overly complicated, while another trooper found this particularly true of the dismounted tactics. The adjutant of the Tenth Cavalry questioned some of the fine points of the work, just as some infantry officers had done with their volume. As for the artillery tactics, in 1895 one lieutenant pointed to the shortcoming that they lacked a training course for gunners.[6]

These criticisms grew to the point that in 1894 the commanding general of the army, John Schofield, wanted each of the three Leavenworth manuals reevaluated and, if necessary, replaced.[7] Secretary of War Daniel S. Lamont turned to General Ruger, the initial reviewer of the 1891 volumes, to oversee this project.[8] Two new boards of officers studied the artillery and cavalry tactics and General Ruger reviewed their efforts. Ruger

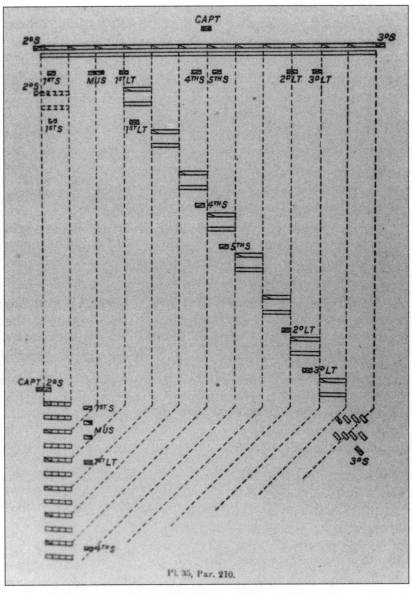

A plate from the 1891 infantry manual, illustrating a company
moving from column to line. Troops needed long hours of
drill to master movements like this one.

himself, with several aides, spent long hours reassessing the infantry volume.[9]

A two-member board, consisting of Major Henry C. Hasbrouck and Captain James M. Lancaster, reexamined the 1891 artillery manual, finishing the task "promptly and well" by the end of 1894.[10] General Ruger added some revisions, Secretary Lamont endorsed the results in February 1896, and the War Department published this work as the army's authorized artillery tactics.[11] The 1896 volume remained much the same as the 1891, with an expanded "School of the Soldier Dismounted,"[12] an abbreviated "Manual of the Sabre,"[13] and other slight changes.

The revision of the cavalry tactics paralleled that of the artillery. While Hasbrouck and Lancaster were studying the manual for the field guns, Lieutenant Colonel Louis H. Carpenter and Captain Edward S. Godfrey of the Seventh Cavalry and two other cavalry officers met at Fort Riley, Kansas, to review the volume for the mounted arm.[14] General Ruger revised their work, Secretary Lamont again endorsed the results, and in 1896 the army replaced its 1891 cavalry manual.[15] This work was even less innovative than the 1896 artillery tactics. The cavalry reviewers added a few "general principles" to the close-order drill[16] and made slight alterations in the extended-order section[17] but otherwise left the 1891 system in place. Working with Captain William C. Brown, Lieutenant Colonel Carpenter proposed some additional revisions to the 1896 volume, but the War Department failed to endorse these recommendations, and so the cavalry entered the Spanish-American War with essentially the tactics it had had since 1891.[18]

The work on the infantry volume took far longer than the cavalry or artillery. General Ruger, who had never been enthusiastic about the 1891 foot tactics, threw himself into their revision. He and his aides scrutinized the infantry manual, spending the year between mid-1895 and mid-1896 on its close-order section alone.[19] In February 1897 Ruger sent his draft of this part of the volume to Lieutenant General Nelson A. Miles, who had succeeded Schofield as commanding general in October 1895, allowing the senior officer to review it while Ruger continued to work on the open-order section. When the spring of 1897 arrived, almost two years had been spent on these revisions, and doubtless many officers wondered if anything would ever come of the project. One soldier noticed that the artillery and cavalry were "now forging ahead out of sight" with their 1896 manuals and asked if the infantry would "be given a little of the same dose."[20]

The War Department might have given this question an affirmative answer and the infantry, like the artillery and cavalry, might have gained a new manual in the mid-1890s, if Ruger had completed his work before Miles became commanding general. The project lost momentum because the army's new senior officer took a personal interest in tactics and closely studied his subordinate's work. Miles was troubled that the draft provided solely for a three-battalion regiment at a time when the single-battalion regiment remained the authorized organization. He had endorsed the three-battalion scheme in 1895 but changed his position the following year. Drawing on his own Civil War experience, Miles recalled that the ten-company, single-battalion regiment had proven effective during that severe conflict, and he felt no urgent need to change the form of organization. The commanding general may have rejected Ruger's draft because of this issue alone, or perhaps he had additional objections. As one informed observer put it: "General Miles is known to have decided ideas of his own on the subject of drill regulations."[21]

While some officers like Miles were interested in improving the infantry tactics, others embraced the conservative spirit that persisted in the army throughout the 1880s and 1890s. "Military men of all others are the most averse to change," observed Major Eben Swift in despair, "and the slowest to accept new ideas." A captain stationed at Fort Bliss, Texas, wrote in 1888: "The Soldiers of the United States are accustomed to Upton's Tactics and any great departure from them would be to learn a new system—new commands and new movements—and the unlearning of a system now known and in use, tactics which we have not tried in war and do not certainly know whether they will serve present requirements." "I know there is a large class of officers," another soldier confidently declared in 1894, "who object to change on general principles, urging that the present order of things is good enough for peace, and that in war we will learn again as we did in the last. We would learn, but too late."[22]

During the 1880s some theorists remained content with Upton's system of fours and others even continued to endorse linear formations similar to those used during the Civil War. The infantry manual that William H. Morris offered the army in 1882 provided for troops to march in fours, maneuver into battle in two lines, and fight in single rank with a second in support. Three years later, another regular officer advocated sending regiments into combat in a two-line formation and urged that, even in attacks against the new weaponry, the supporting line must follow quite closely the front one. In his 1889 *The Campaign of Königgrätz*, Arthur Wagner criticized the Prussians for their small columns and favored instead the traditional formation in successive lines, the standard deployment of the American Civil War.[23]

As late as the 1890s, some officers continued to resist loose-order formations. In 1891 Colonel Thomas M. Anderson pointed to the case of soldiers behind a parapet, who would "surely be put shoulder to shoulder" and then added that "even in the open field, the question will often arise, how can the greatest loss be inflicted on the enemy[?]" A year later Captain W. V. Richards argued that an extended formation hindered an officer's ability to control his men and reduced the chances of driving an attack home. He contended that a loose line could never assault a defender in "an overwhelming mass. We seldom hear of a skirmish line carrying a defended position." Major Joseph P. Sanger, an inspector-general officer, offered the opinion that the French and Germans were becoming disenchanted with loose order and speculated that they would revert to close-ordered formations much like those of the Union and Confederate armies. His superior Joseph C. Breckinridge, also a veteran of the Civil War, agreed and endorsed the time-worn deployment into successive lines as the assault formation that benefited from the greatest firing front, while offering defenders the smallest target. One volunteer soldier recalled the traditional training that his regiment received after the outbreak of the Spanish-American War. "Those were the days of close order," he reminisced, "and we drilled in it much as in the days of Waterloo or Gettysburg."[24]

A few officers went beyond the defense of conventional ideas, disparaging any study of tactics. Captain James Chester, one of the most hidebound traditionalists, contended in 1886: "I believe it would be a blessing to our profession if the whole literature of the art of war was destroyed. Young minds are so easily poisoned by the dogmas of the text-books, and such learning is so hard to get rid of." Ten years later, Chester argued that while officers might master their drill books, they would "never understand" battle tactics. "Better stick to parade maneuvers and sham battles," he advised his fellow professionals, "and [the] country, if it knew the facts, might devoutly pray that it may never have any other occasion for your services." An 1885 correspondent to the *Army and Navy Journal* also discouraged the study of tactics. "It seems to me that so far as our tactics are concerned things are getting abnormally mixed," he wrote, "and the more we study the worse we tangle." The same year, another veteran claimed that many older officers made a virtue of their ignorance of tactics and bragged to their younger comrades about how little they knew of the subject.[25]

Such extremists probably were few in number, but a sizable group of officers gave little thought to tactics during the 1880s and 1890s because they believed that no innovations were needed to defeat their Indian opponents and that foreign wars were unlikely. "I see no probability," one ordnance officer declared in 1881, "of the employment of our cavalry against any large or well organized bodies of civilized cavalry." The commanding general of the army, Lieutenant General Philip H. Sheridan, expanded on this point in his official report of 1884. "Excepting for our ocean commerce and our seaboard cities," General Sheridan advised, "I do not think we should be much alarmed about the probability of wars with foreign powers, since it would require more than a million and a half of men to make a campaign on land against us." Three years later he developed a similar argument at a public banquet in Philadelphia. Doubtless Sheridan based these remarks on an Office of Naval Intelligence study that had concluded that it would require about three-quarters of all the shipping of Europe to bring a large army to the American coast.[26]

The day-to-day demands on soldiers during the 1880s and 1890s left them with little time for tactical study, field maneuvers, or even parade-ground drills. In 1886 the secretary of war complained that once a second lieutenant had secured his commission, he ended his military studies. Colonel Alexander McCook, a Civil War veteran stationed at Fort Fred Steele during the 1880s, became frustrated that he had "but little chance to do soldier work." "There is no telling," despaired the soldier-scholar

Matthew F. Steele in 1895, "what duty an army officer may be put at next." The time available for field maneuvers was limited, even at the Infantry and Cavalry School. "The practical training of the officers here is not carried to the desired extent . . . ," Colonel E. S. Otis acknowledged in 1884. "They do not participate in, and become practically acquainted with, the field service of troops, nor the varied maneuvers in which they engage in actual warfare, which depend upon features of country, character of troops and armament, and which come within the domain of strategy and grand tactics." Otis went on to explain that the lack of time for field maneuvers was due to "the continued employment of men in duties connected with the post." Some officers were troubled that even the most basic drills deserved more attention than they received. One cavalry lieutenant expressed concern in 1893 that a five-year enlistment was too short for a trooper to master his required drills, given the demands on his time at most posts. Major General Oliver O. Howard urged in the mid-1880s that company and battalion drills should be as mandatory as target practice. "There is no cogent reason," he insisted, "why every enlisted man, except one cook with each company, should not be required to give one hour out of the twenty-four to his profession."[27]

The cavalry in particular had difficulty finding time for tactical training. "The old strain is still kept on the cavalryman," lamented Major Edwin V. Sumner in 1888, "and he is expected to do everything about the post in the way of guard, fatigue, etc., that an infantryman does." Guy V. Henry, a cavalryman, once raised this same complaint with an infantry colonel and post commander, who answered that "this extra work was fully compensated for by the privilege of having a horse to ride."[28] Captain William H. Carter believed that much "of the unsatisfactory condition" of the cavalry's tactical training during the mid-1890s was "due to the fact that the maintenance of a lot of frontier villages—misnamed forts—falls to the lot of cavalry commands."[29]

The Indian wars demanded the attention not only of the cavalry but of the entire army, discouraging its officers from studying the tactics that they might need against a European opponent. Brigadier General John Pope contended in 1881 that Indian campaigning was not "conducive to the proper discharge of military duty or the acquirement, either in theory or practice, by officers or soldiers, of professional knowledge or even of the ordinary tactics of a battalion." Long after the Wounded Knee tragedy of 1890, historians identified that heart-rending episode as the closing event of the Indian wars, but contemporary army officers did not see it that way at all. They did not perceive Wounded Knee to be an Appomattox or Guadalupe Hidalgo, marking a final, decisive end to a well-defined war. They understood the Indian wars as a long-running, open-ended conflict, without a clearly marked beginning or end. The army devoted a great deal of attention to Indian affairs, into the twentieth century, and its officers in 1890 had no way of knowing when this commitment would end.[30]

This continuing responsibility on the frontier diverted soldiers from drill and tactics, and the small size of the army posed another hindrance. Between 1880 and 1897, the army's total strength never exceeded 28,500. In 1883 it had only 25,652 officers and enlisted men, scattered across more than 115 posts. Commanding General Philip H. Sheridan complained in 1887 that the enlisted strength of the army's companies was too low to conduct satisfactory exercises, maneuvers, or even basic drills. Major General John M. Schofield, then commanding the Division of the Atlantic, reinforced his superior's point of view, in a report prepared the same year. "The present small number of artillery troops," Schofield found, "can not even be provided with the means necessary for their practical instruction." "It has not been possible to concentrate garrisons," Brigadier General E. S. Otis reported from the Department of the Columbia in 1896, "and hence the smaller ones have been obliged ... to feign conditions, circumstances, and obstructions by a very active cooperation of the imagination."[31]

General Otis's comment pointed up the fact that the army was not only small but also scattered. Its forces were isolated at small-unit posts, making large-scale field maneuvers costly, and therefore rare, events. The long-tenured veteran Brigadier General John R. Brooke noted in 1890 that the army had not assembled a full division since the Civil War and urged the commanding general to do so. Brooke recommended sending companies and regiments into the field to participate in a division-sized "Camp of Instruction," leaving behind detachments to look after their posts. The commanding general, Major General John M. Schofield, considered this proposal but determined against holding any large field maneuvers in 1890, to save money. Major General Thomas H. Ruger complained in 1896 about the limited funds available to assemble infantry for field exercises, and at about the same time, the army's inspector general cataloged the obstacles to infantry drills: "a defective organization, the restricted area of many military reservations, and the lack of frequent aggragation [*sic*] of a considerable force."[32]

Inadequate field training also hindered the cavalry. Captain Frank K. Ward, who graduated from West Point a year after Appomattox and spent a long career with the mounted arm, admitted in 1893 that he had never seen a full regiment of cavalry and had "no practical knowledge of regimental movements." Four years later another mounted officer complained that, with rare exceptions, the troopers did not conduct practical field maneuvers. That same year another cavalryman bitterly speculated: "To suggest that brigade evolutions be had each year would . . . bring down the wrath of all good Congressmen upon my head. We can sink a million dollars in some little canal away down in Florida, but when it comes to spending a few thousands to increase the efficiency of our army—well, that is a different question."[33]

The artillery, and at least two of the army's schools, struggled with the same problem. Tasker H. Bliss, a first lieutenant in the late 1880s who later became the army's senior officer, saw how hard it was for field gunners to train properly when their batteries were so widely scattered. Bliss proposed to remedy this defect by assigning five batteries to a single post and two batteries to each of two others. Artillerymen at these three consolidated stations would be able to practice "the higher tactics of the arm, . . . the ready and skilful handling of large masses of field batteries upon which the fate of battles will depend." Small companies hindered the tactical exercises at the Infantry and Cavalry School. "Five or six officers cannot be advantageously employed at an infantry or cavalry company drill," E. S. Otis complained in 1883, "when but two or three sets of fours participate. . . . [Tactical instruction] cannot be satisfactorily accomplished if drills revert to farces on account of insufficient strength in organizations." As late as 1888, Lieutenant John P. Wisser of the First Artillery faulted the Infantry and Cavalry School, and also the Artillery School, for devoting too much time to teaching theory and too little to practicing in the field.[34]

Soldiers assigned to these small units lived an isolated, dreary existence, and some observers believed that this expanded the group of moss-back officers who had neither interest in, nor energy for, tactical exercises and who stubbornly resisted new ideas. Brigadier General John J. Coppinger, who put in a long career spanning four decades including service with four different infantry regiments, concluded that military life on isolated army posts made some soldiers more closed minded. Late in 1890s, he observed that a number of soldiers did not want to discuss the tactical decisions they made during field exercises, betraying "an extreme sensitiveness to criticisms which . . . isolation has developed in some officers." Frederic Remington became famous for his paintings of the Indian wars and, like General Coppinger, he was a keen observer of military life on the frontier. In 1893, he offered a harsh opinion of the army's veteran leaders: "The old gentlemen who have taken care enough of their healths to get into the high grades . . . are responsible for the dry rot which is disintegrating our army to-day. It will probably take a war to make them resign. . . . There are plenty of infantry field officers who couldn't get a pony past a dead dog in the road if they were following their regiments in action, not to dream of six rails on a snake fence."[35]

Remington was speaking of infantry officers, but the other arms of the service suffered from the same isolation and stagnation. "I despair of ever getting a command large enough, even for amusement, in this Department," cavalryman Edwin Vose Sumner wrote from Fort Robinson, Nebraska, in 1883, "and I have no words to express to you the monotony of service year after year, at a post like this, with 'a corporals guard.'" Several years later, Sumner reflected that the isolated, lethargic conditions continued at some posts year after year, undisturbed by the annual visits of the inspector general. A lieutenant of the Tenth Cavalry complained that no training was done at Fort Custer, Montana, during the winter of 1892–1893: "For months the horses have been tied up to the picket or manger. The puerile excuse is given that it is too cold." Henry L. Harris, regimental adjutant of the First Artillery, longed in 1889 for the day when "every non-commissioned officer should be thoroughly informed theoretically & practically in *Artillery*. It will take time, and many stiff necked imbeciles will have to be suppressed; but I have faith in the future, though at times I am compelled to say 'How long, Oh Lord! how long?'."[36]

The small size of the units available for exercises at most posts stagnated the study of tactics, and so did the modest land area of the stations, which was usually too limited for artillery practice or large-scale infantry maneuvers. One artilleryman reminisced in 1896 that while light batteries had once found it easy to practice, now it was difficult to find safe places for firing. Infantrymen faced a similar dilemma: at a time when improved weapons put a premium on loose formations, many posts lacked enough space for training in extended order. This problem prevailed, Major General Oliver O. Howard reported in 1894, throughout the populous Department of the East. Colonel Robert P. Hughes toured the region a year later and echoed Howard's concern: "The reservations are so limited in area that instruction of value in extended order is not possible." Hughes proposed correcting "this very serious deficiency . . . by hiring territory at some central point which can be reached by several commands by reasonable marches, and having a well-matured course of instruction in field work." The population of the West was sparser than that of the East, but the acreage available for tactical maneuvers was often equally limited. On the eve of the Spanish-American War, neither Fort Meade, South Dakota, nor Fort Crook, Nebraska, had enough land to accommodate large-scale exercises, and the farmers south of Omaha lodged complaints when troops from the latter post forayed onto their property.[37]

Even when posts were able to conduct realistic field maneuvers, the results sometimes proved disappointing. One artillery lieutenant complained in 1888 that, in place of "practical work," many posts still gave their time to "drill and target practice." One cavalry officer reflected in 1897 on his seven years of assignments: "At nearly every post where I have served, very little practical instruction was given in minor tactics. Once or twice a year we were ordered out and made an attempt to post an outpost or to act as an advance guard, but the work was always very poor." After one particularly frustrating ride of instruction conducted in 1891, Major John

Breckinridge Babcock wrote that the commanding officer failed to screen his advance from enemy scouts, "spent hours" advancing up a ridge through "thick and almost impenetrable brush," and left both of his flanks in the air. "The whole performance made me 'very tired,'" Babcock recalled, "and was quite the reverse of instructive to the students." General Coppinger acknowledged that the battle exercises held in his department during 1895 had fallen "short of a practical representation of actual combats." Guy V. Henry criticized a field maneuver staged by two of his own troops from Fort Ethan Allen, Vermont, in 1897. The ever-skeptical James Chester concluded that "of all the shams in existence, the sham battle is the most absurd. It lacks the only element which constitutes a battle, and is therefore a fraud and a teacher of falsehood."[38]

These limitations on exercises of the regular army also hindered the volunteers and militia, who struggled with additional problems of their own. One member of the Seventy-first New York Volunteers described the regiment's training for the Spanish War as "nice, formal, punctilious, and with every move to the singsong of a bugle. It was half concert and half fight." Most regular officers looked askance at the training conducted by the National Guard, a fledgling organization in the 1880s. Inspector Charles King reviewed the Guard units of Wisconsin that decade and found a tactical hodgepodge. Some of the state's companies practiced Civil War tactics; others, with men of German heritage, used Prussian drill; and still others allowed their captains to make up systems of their own. Captain William H. Powell, a Fourth Infantry officer who had won brevets at Antietam and Petersburg, took a jaundiced view of a sham battle waged by militia units at an 1884 encampment. The veteran regular criticized the National Guard commanders, who, because they wanted to stage a lively show for their audience, formed and maneuvered their troops in the open rather than keeping them under cover as they would in combat. Trying to please the Dubuque, Iowa, crowd, the attackers took up a position that, Powell jeered, "would be simply ridiculous in time of war." Captain George B. Rodney of the Fourth Artillery inspected two brigades of the Massachusetts Militia during 1882 and commented: "The First Brigade was exercised in a sham fight, the time expended in this work being time wasted." Career soldiers also resented the fact that many spectators regarded militia drill competitions as festive sporting events and enjoyed gambling on them, ignoring their importance as military exercises. Dismayed by these contests for "large money prizes," where "pools [were] sold upon the results, as at a horse race or game of base ball," two regular officers recommended that their fellow professionals "be discouraged from acting as judges or referees in military contests where sums of money are given as prizes or hazarded on the result."[39]

Referees encountered a number of other thorny problems, including the responsibility for declaring the winner of a field exercise. "The battle of Maine's Pass was fought this morning," one newspaper reported of an 1895 New York National Guard contest, "but it will be a little difficult for the referee to decide who won. Col. Francis V. Greene's forces got around on the left flank of Guy V. Henry's command, but whether they could have climbed the steep hill under Gen. Henry's rapid fire is a question that bullets alone could conclusively decide." An officer at Fort Riley acknowledged that when an exercise neared its "crisis," "nothing short of real battle and bloodshed can decide the question of victory or defeat." Mounted charges were particularly hard to adjudicate. The rules in place in 1897 at Fort Riley provided: "Umpires should be early on the scene in cases of cavalry attack, as otherwise it might be difficult to judge." A few years earlier, the *Journal of the United States Cavalry Association* quoted a similar stricture from the German field service regulations.[40]

The rules for field maneuvers at Fort Leavenworth's Infantry and Cavalry School declared that an umpire's decision "must be . . . at once obeyed,"[41] but commanders of competing units could be quick to challenge a referee's ruling. Army officers were competitive men who resisted defeat in battle, whether the combat was real or an exercise. Like the managers and coaches of twentieth-century athletic teams, they viewed the officials of their contests with wary skepticism. Doubtless many American cavalry officers joined in the complaint of a British journal that referees tended to rule mounted charges unsuccessful, no matter how rapidly or cleverly the horsemen marshaled their assault. Colonel James W. Forsyth, commandant of the Cavalry and Light Artillery School, doubted that any umpire could "demonstrate to the satisfaction of any force that its opponent has gained a victory."[42]

Moreover, as Colonel Forsyth also noted, exercises were held primarily to practice tactical principles and not as games between two commanders. The chief problem of umpires was not determining the "winner" and "loser" of a field competition but helping the participants learn as much as possible from the exercise while keeping it running smoothly. Referees tried to balance two considerations: on the one hand, the soldiers might learn a great deal if an umpire frequently stopped an exercise and allowed the men and officers to evaluate the positions of their units; on the other hand, this style of refereeing disrupted the flow of a maneuver and diminished its realism. Forsyth himself studied thoroughly the umpiring in American and European armies but failed to find a satisfactory system.[43]

The challenges of conducting realistic training and developing innovative tactics were compounded by the technological revolution that affected all of late nineteenth-century America. Eclectic inventors bombarded the country's military leaders with a variety of contrivances, some of which proved valuable, others not so. During the 1890s a Rhode Island businessman designed a set of entrenching tools that he was confident the army would want to purchase, while a Chicago firm offered "the Austin Steel Reversable Road Machine," which would save soldiers the labor of digging breastworks. A Mississippi entrepreneur was equally certain the cavalry needed his ring-strap-and-catch device, which would enable a single rider to lead as many as twelve horses. William D. Riley offered the army a scheme as wild as any, his "plan to Armour in Coats of Mail, one Thousand Men, the Coats to be manufactured out of Aluminum." The imaginative Philadelphian urged this idea upon General Schofield in 1891, warning the American commanding general that a European rival might use this new technology to decisive advantage. "I see that the Emperor of Prussia has made a large Contract in Pittsburg, in this State, for Aluminum . . . ," Riley darkly hinted, "and it is my opinion that he intends Placing Coats of Mail on Some of the Branches of the Army[;] if he does, he will Walk through any thing he comes against, that [is] not so Protected. The Prussians are close Mouthed and know how to keep a Secret, to which may be attributed their Success in Defeating the Austrians and French with the Ne[e]dle Gun."[44]

Many of the new inventions proved far more practical than William D.

Riley's coats of armor. The dynamite gun, which tough-minded soldiers at first may have thought a farfetched idea, saw field service during the Cuban insurrection and later the Spanish-American War.[45] Breech-loading repeaters, Gatling guns, breech-loading field artillery, sophisticated entrenchments, metallic cartridges, fixed ammunition, smokeless powder, and other technological advances revolutionized the nature of warfare during the late nineteenth century.[46]

Improved weapons made loose order a desirable formation, but the farther apart that infantrymen were deployed, the more difficulty their officers had controlling them. During the 1880s and 1890s, commanders still had no field radios and so extended deployments remained impractical. Nor was it possible to sustain a skirmish formation, after it began suffering casualties to rapid-firing weapons, without some accompanying confusion and loss of order among the surviving, and supporting, troops. Francis V. Greene, one of the army's most dedicated students of military history and theory, analyzed this predicament in an 1883 article for a professional journal. Contemporary firearms and entrenchments, he pointed out, forced attackers to advance in a thin line, continually replacing their losses with reinforcements from the rear. This inevitably meant mixing together men from different commands, and however long the front or whatever formation used, there was "no possible way to avoid . . . elements of confusion." At about the same time as Greene's article, another soldier counseled: "There appears to be no way in which this constant feeding of a skirmish line from the successive fractions of a column can be obtained, while still preserving the relative places of the men."[47]

The loose-order dilemma exemplified the problems that made the 1880s and 1890s such daunting decades in the tactical history of the army. Several obstacles hindered the development of battlefield doctrine. Rapid changes in technology made it difficult, if not impossible, to restore the balance between offense and defense. At the same time, the army's small size and sparse resources restricted its ability to test tactical ideas, and disagreements over fundamental issues of battlefield theory prevented the service from reaching a consensus about its tactics.

These problems defied any permanent solution, as had always been the case with tactics. Ever since ancient times, advances in weapons had demanded changes in methods of fighting and the development of tactics had been an ongoing process over the centuries. This historical pattern became more apparent to military theorists than ever before during the late nineteenth century, when a technological revolution dramatically accelerated the rate of weapons improvements.

These dynamic changes made officers more aware than they had ever been in the past that successful tactics now depended on flexibility and innovation rather than dogma. The officers who had prepared drill books during the eighteenth and most of the nineteenth centuries could be quite confident that the formations and movements in their manuals would serve long into the future. By the late nineteenth century, theorists had become more aware of the impermanence of their work and, for the first time, they acknowledged without prompting that the army would never arrive at a "final" tactics. In 1891 Colonel Henry W. Closson of the Fourth Artillery offered a metaphor that blended the Old Testament's Isaiah and the army's Upton: "No tactics can ever be a finality until the lion and the lamb, the little child and the leopard, form a set of fours." Francis Greene argued early in the 1880s that tactics had become a matter of "the genius and instinct of the commanders. No books can teach this, and no rules define it." Another soldier expanded on Greene's point: "The many variations of ground and the problems which war presents make it impossible to construct a theory which will cover every situation likely to occur therein. . . . the best practical theory can only serve as a most general support." "The gravity of the duty of preparing drill regulations, and the labor attaching to it," the pragmatic William H. Carter observed, "are not always appreciated by those who criticise. It is not always a question of what is best, but what is most expedient and applicable."[48]

For the first time in the army's history, some of its officers acknowledged that future battlefields might pose problems that no tactical theory could resolve. A lieutenant of the Eleventh Infantry identified one example in 1897: the choice between close and open formations. While dense attacks would be annihilated, loose ones would be difficult to control. "Choose your own horn of the dilemma," this infantry officer despaired, "one seems quite as bad as the other." Thomas M. Anderson, the colonel of the Fourteenth Infantry, grimly acknowledged another example, perhaps the most daunting of the army's combat problems. Reflecting on the rapid and accurate fire of the weapons of the 1890s, Anderson starkly predicted: "The loss [in battle] will be severe with any system of tactics that can be devised."[49]

8

*Charging against
Entrenchments
and Modern Rifles*

The Spanish-American War
and the First Phase of the
Philippine War, 1898–1899

When Congress declared war on Spain in April 1898, the United States had not waged a conventional war for thirty-three years and had not fought against a foreign opponent for half a century. The American army had adopted new tactics during the 1890s, but it had not given them the crucial test of combat. As one year of peace followed another, some officers became concerned about their army's lack of practical experience. The devoted artillery student William Birkhimer worried most about his own arm. "There has been no war to test the armaments of the present day," he noted in 1885, "particularly those of artillery." On the eve of the Spanish-American War, a cavalry officer warned of a larger danger: "We are gradually becoming more involved in the affairs of other nations. . . . War can be declared, begun and finished today in less time than was required to place our unwieldy armies in their first positions in 1861." Another cavalryman, William H. Carter, acknowledged in 1900 what virtually every American officer must have recognized in April 1898, that the nation had not been prepared for war.[1]

The United States defeated Spain quickly and relatively cheaply, and most Americans soon forgot that when Congress had declared war the outcome had been uncertain and that the Spanish army in fact had entered the conflict with some significant advantages, including an ability to deploy large numbers of troops around the world. Spain sent 150,000 regular troops to suppress the Cuban rebellion, a force that although weakened by poor training, tropical diseases, and dispersed deployment was impressive in sheer size. It was far larger than the command that in 1864 Ulysses S. Grant had led against Robert E. Lee or William T. Sherman against Joseph E. Johnston. No American general directed so large a force until World War I. While Spain had been able to send 150,000 regulars to Cuba, the entire United States Army consisted on the eve of the Spanish War of only about 28,000 troops.[2]

This advantage in numbers carried over to the individual campaigns of the war. For its attack against Manila, Major General Wesley Merritt's Eighth Corps deployed 8,500 troops,[3] while Governor-General Basilio Augustín had a slightly larger force defending the city, and his total strength throughout the Philippines, regulars and militia, came to about 40,000. Had the Spanish concentrated their troops, they would have outnumbered the Americans by more than four to one.[4] During the Santiago campaign, General Arsenio Linares commanded about 36,000 troops across eastern Cuba, more than double the 17,000 of Major General William R. Shafter's Fifth Corps. The Spanish again dispersed their forces and in this case they lost the advantage of their numbers, with only about 12,000 soldiers holding Santiago and its outposts against the 17,000 Americans.[5] As for the Puerto Rican campaign, the Americans fielded a larger force than their opponents, but the difference in absolute numbers was so slight that the Spanish defenders held a relative advantage.[6]

The Spaniards were also in some ways better armed than their opponents. While the Spanish infantry carried clip-fed, bolt-action Mauser rifles that fired smokeless cartridges, the American volunteers fired single-shot, breech-loading .45-caliber Springfields that used charcoal black powder.[7] The Springfield's enormous stopping power was its only strong point. "There was one thing to be said for those old Springfields that the volunteer troops were armed with," Brigadier General Frederick Funston conceded, "and that was that if a bullet from one of them hit a man he never mistook it for a mosquito bite." Having acknowledged this weapon's sole advantage, Funston lamented that the Americans continued to use it, while the Spanish carried the longer-range Mauser, which he rated a splendid shoulder arm. Private Charles Johnson Post wistfully described the Mausers as "excellent weapons"[8] and complained that the Springfields produced "a cloud of white smoke somewhat the size of a cow." The Springfield had a lower muzzle velocity and higher trajectory than the Mauser, making the American weapon harder to aim at long ranges. At 1,000 yards the bullets of a Springfield came down at a considerable angle and reduced, as General Funston pointed out, using a phrase familiar to his generation, "the dangerous space."[9] The Springfield was also difficult to handle. It could knock down two soldiers, the opinionated Private Post claimed facetiously, the victim it hit and the man who fired it. The "old Springfield," Major General Joseph C. Breckinridge pronounced, "seemed a begrimed and suicidal blunderbuss upon the battlefield."[10]

The American regulars carried the Krag-Jorgensen, which like the Mauser used smokeless powder and was greatly preferred over the Springfield. One soldier whose regiment received Krags after the Spanish War wrote home in delight: "We have had the new up to date rifle, Krag-Jorgensen, issued to us since I wrote to you last. . . . The rifles are beauties. Although you have to keep cleaning them every day or two they are easy to clean." This infantryman's enthusiasm was merited but represented only part of the story: the Ordnance Department had trouble keeping the weapon, and its ammunition, in supply throughout the Spanish-American War.[11]

Smokeless powder benefited the Spanish as early as the opening skirmish of the war, at Las Guasimas on June 24, 1898. Brigadier General S. B. M. Young reported that in this action his men could judge the position of the enemy only by the sound of his fire. The well-known war correspondent Richard Harding Davis described the same encounter: "[The Spanish] bullets came from every side. Their invisible smoke helped to keep their hiding-places secret, and in the incessant shriek of shrapnel and the spit of the Mausers, it was difficult to locate the reports of their rifles."[12] Another famous participant, Theodore Roosevelt, then lieutenant colonel of the First United States Volunteer Cavalry, the Rough Riders, gave a similar testimony. "The effect of the smokeless powder," the future president wrote, "was remarkable. . . . as we advanced we were, of course, exposed, and they could see us and fire. But they themselves were entirely invisible." "The Spaniards used smokeless powder," Brigadier General Francis V. Greene reported of a fire fight near Manila, "the thickets obscured the flash of their guns, and the sound of a Mauser bullet penetrating a bamboo pole is very similar to the crack of the rifle itself."[13]

In addition to smokeless powder, terrain favored the Spanish. Fighting virtually the entire war on the defensive, they were able to take advantage of the jungles, rivers, poor roads, and other obstacles to American operations in Cuba, Puerto Rico, and the Philippines. General William R. Shafter pointed out that some stretches of the road approaching the San Juan Heights were too narrow for his men to march in column of fours, and the undergrowth alongside the route was too dense for skirmishers. This Cuban terrain was not an exceptional case; the Americans contended against difficult ground throughout the brief war. In a telegram to his superiors, Major General James H. Wilson tersely described what the soldiers of the First Corps found in Puerto Rico: "Country very rough indeed." Recounting a trek through the island's Rio Seco Valley in August 1898, Brigadier General Peter C. Hains reported: "We struck across the country, ascending the mountains by a path that was scarcely passable for horsemen in single file. . . . A part of the march was made over one of the roughest of mountain roads, and the distance covered was extraordinary, considering the heat and climate of this country." During the operations against Manila, Brigadier General Arthur MacArthur, father of the famous World War II and Korean War commander, led his brigade across challenging terrain. MacArthur's advance, as another officer recounted it, was "hampered and intersected in all directions by swamps and paddy fields; the brush was thick, and the enemy's line particularly strong at this point."[14]

American soldiers also struggled against difficult terrain during the Philippine War, which followed on the heels of the conflict with Spain. Brigadier General S. B. M. Young referred to the region north of Manila in 1899 as a "bottomless pit." Another brigadier, Frederick Funston, recalled at least two occasions during the Philippine War when the Americans were embarrassed by their ignorance of the terrain. "The country on the front of [Brigadier General Irving] Hale's brigade was to us a veritable unknown land.... It turned out to be a dense tangle of forest and undergrowth, cut up with ravines." The second instance, involving an attack by Funston's own troops, provided a more dramatic example. "The men had raised the usual yell," Funston remembered, "and we thought that in a few seconds we would be among [the enemy], when we were brought up with a start, and the men instinctively threw themselves on the ground. We had rushed to the very brink of a river about eighty feet wide and ten feet deep."[15]

The Americans faced enough difficulties in 1898 that some officers were wary about predicting the outcome of an encounter with Spain, their professional caution running counter to the public's war fever. The month before the conflict began, the hardened veteran Edward S. Godfrey warned against overconfidence. "Lots of people think we could clean out any invading army with pitchforks," Godfrey believed, "and that in any operations against Spanish forces they think all we have to do is hold the bag and the Spaniards will run into it! The Spaniards may not be very enterprising but they are not cowards." An American officer who made the attack on the San Juan Heights wrote shortly after the battle: "We are willing to concede that [the Spanish] are good soldiers, have good arms and do good shooting, as is attested by the number of our casualties." Frederick Funston held similar views about the revolutionaries who opposed the Americans in the Philippine War. Although he held a low opinion of the marksmanship and fire discipline of the Filipinos,[16] the Kansan considered them worthy foes. They were not intrepid attackers, Funston judged, but they were staunch defenders who would hold a trench until it was lined with their dead.[17]

The United States defeated Spain rapidly, and the victors who had expressed any prewar reservations no doubt quickly forgot that they had ever entertained them. "The taking of Guam was a comedy," acknowledged one American soldier. "The taking of Manila by our troops was also somewhat of a farce." Brigadier General Arthur MacArthur was less disparaging about the attack on the Philippine capital but conceded that it could not be considered a major military event. The two brigades that captured Manila on August 13 lost only fifty officers and men.[18] The Puerto Rican campaign consisted largely of marches rather than battles. Richard Harding Davis, a participant, called it a "picnic"; the Americans took even fewer casualties than they did while entering Manila. The battles of the Santiago campaign, El Caney and San Juan, were the only actions of the conflict that produced sizable American losses. As Frederick Funston pointed out, the undeclared Philippine War proved to be a longer and much bloodier struggle than the brief Spanish War waged before it.[19]

Although the battles of the Spanish-American War were small, they served to reinforce many of the army's prewar ideas. The limited combat experience of 1898 seemed, for example, to verify the prediction that small units would assume greater responsibilities in future operations. Major General Nelson A. Miles delegated the conduct of the Puerto Rican campaign to four minor columns, which, starting from dispersed points around the southern coast of the island, moved along separate lines into its interior.[20] A company officer of the Twenty-third Infantry described a typical small-unit action during the operations against Manila. "[The Spanish fire] was individual firing but rapid," Second Lieutenant Celwyn E. Hampton said of a brush with the enemy on August 5, "and indicated a considerable force. . . . I returned the fire, the men firing by squads and at times at will." A captain described the famous advance against the San Juan Heights: "The fight, as far as I could see, was carried on by companies acting independently."[21]

As small-unit actions gained in importance, it followed that junior officers would assume larger responsibilities in combat. This prewar prediction was realized as early as the opening skirmish at Las Guasimas, where the company officers of the Rough Riders were forced to act on their own because, as Richard Harding Davis pointed out, the difficult Cuban terrain prevented Colonel Leonard Wood or Lieutenant Colonel Theodore Roosevelt from controlling the regiment. Wood and Roosevelt gave their captains "general orders," Davis asserted, and then relied on "the latter's intelligence to pull them through. I do not suppose Wood saw more than thirty of his men out of the five hundred engaged at any one time."[22] During the famous attack on San Juan Hill, Captain Leven C. Allen, the commander of the Sixteenth Infantry's Company C, struggled through the barbed-wire fence in front of the Spanish position. A few weeks after the charge, he recalled what happened next: "I observed that others of my regiment [had cleared the fence and] were out in the field, Captains [Charles H.] Noble, [George H.] Palmer, [William C.] McFarland, and [William] Lassiter, each leading his men to the front. I dashed forward, followed by my company, and we five captains led the charge." Junior officers paid a price for their new role: in most of the regiments that attacked El Caney and San Juan, the officer casualty rate was about double that of the enlisted men.[23]

One striking example of a company officer leading an independent operation took place during the attack on Manila. Captain Stephen O'Connor separated his fifteen skirmishers from their regiment and brigade, found an unopposed route into the city, added a few stray American infantrymen to his small command, did some street skirmishing, and seized a stone bridge. O'Connor's men doubtless became uneasy when they saw a group of enemy soldiers bearing down on them, but their alert captain quickly sized up the situation. The approaching Spanish wanted only to flee not fight, and O'Connor dispensed with them simply by hurrying them along. Acting entirely on his own, the intrepid company officer had carried off what his division commander called a "remarkable adventure."[24]

The Spanish War also seemed to bear out the expectation that noncommissioned officers, as well as junior officers, would have to assume greater initiative. Brigadier General Arthur MacArthur credited Sergeant Dennis Mahoney with taking a squad on a well-led scout of a Spanish position at Manila,[25] while another, far more junior, officer commended four of his NCOs during the same campaign. Second Lieutenant Lloyd England wrote: "Sergeants [Andrew J.] Gaughron,[26] [Winfield] Harper, [Lindzy E.] Cheatham, and [David J. (?)] Sullivan,[27] who were in command of platoons averaging over 40 men, had perfect control of them." This same lieutenant also praised at length a fifth NCO, a sergeant named Fisher, who led a series of reconnaissance patrols and mapped the difficult terrain that England's battery would have to cross during its advance on Manila. Fisher put this information to good use on the morning of the main attack, when he headed a work party that cut roads and threw bridges over ditches, aiding the advance of the field guns.[28]

The conflict of 1898 also offered evidence of another prewar prediction, that the new emphasis on initiative now extended through the NCO ranks to the individual soldier. The Civil War's blue-clad privates who advanced in close-ordered lines and aligned their ranks by touching elbows had been replaced by infantrymen who fought more independently. Reporting on the battle of San Juan Hill, Captain George H. Palmer of the Sixteenth Infantry identified three members of this new generation of soldiers, Corporal Charles McGiffin and Privates Andrew J. Connors and Frederick J. Liesman. These men, Palmer proudly noted, "were in the first line of the charge [up San Juan Hill] and continued in pursuit of the enemy some distances beyond their trenches and blockhouse, and in front of where our lines were established for the night." Palmer claimed for Liesman the distinction of advancing farther into this part of the Spanish position than any other enlisted man. After General MacArthur's brigade entered Manila, he attributed its success not only to the independent leadership of two company-grade officers, one of them a future army chief of staff, but also to the good conduct of the men in the ranks. He reported that his unit's advance party, led by Lieutenant Peyton C. March and Captain Charles Greene Sawtelle, Jr., had gained a point within less than eighty yards of Blockhouse No. 20, one of the city's defensive strong points. "Aside from conspicuous individual actions in the first rush," MacArthur added, "the well-regulated conduct of this firing line was the marked feature of the contest." Later in the same report, he commended "the work of the soldiers of the first firing line."[29]

A "war story" about one man's exploits during the Manila campaign illustrated how an individual soldier might operate on his own. According to his regimental commander, Private William Beatty of the First Colorado made an independent, intrepid foray in front of the Spanish lines protecting the city. Assigned to outpost duty, Beatty sallied out alone from the American fieldworks and stalked an enemy sharpshooter who had been annoying his regiment. Beatty stripped off some of his uniform, waded a ditch, and warily made his way to a vantage point about 150 yards from the Spanish sniper. The enterprising private then tried to draw his enemy's fire by waving a bunch of grass overhead. The ruse worked—almost too well. The Spaniard put a round into a bamboo stem about a foot and a half from Beatty's head. The Coloradan got off a shot of his own and the sharpshooter slumped over. Beatty fired a second round, and although his opponent remained still, the Spanish infantrymen opened up from along their trenches. Beatty had now used more than his quota of pluck for one day and retreated hastily, wasting no time to confirm that he had killed the sniper. The regimental commander offered his own conclusion: "As Beatty is an experienced hunter and good shot and the range was about 150 yards[,] there is little question that he got his man."[30]

The Spanish War also seemed to verify the prognosis of the proponents of the Gatling gun, when the rapid-firing weapons contributed to the victory at the San Juan Heights and presaged their importance during World War I. Lieutenant John H. Parker wasted no time in capitalizing on this battlefield success. In his official report on the Santiago campaign, Parker was quick to point out that the Americans carried the San Juan works less than nine minutes after his Gatlings opened fire. Immediately after the war he published a history of his detachment's exploits in Cuba, a tract that unabashedly lobbied for the Gatlings. Writing with greater objectivity, American officers from other units also endorsed Parker's weapons. Theodore Roosevelt praised the rapid-firing weapons and their commander, claiming that "Parker deserved more credit than any other one man in the entire campaign. . . . he had the rare good judgment and foresight to see the possibilities of the machine guns, and, thanks to the aid of General Shafter, he was able to organize his battery."[31] In his official report on the campaign, Shafter himself acknowledged the detachment's good service, and the front-line company commanders, whose men benefited most from the rapid-firing weapons, exceeded their corps commander's praise of the rapid-firing weapons. "The Gatling-gun fire," a lieutenant of the Sixth Infantry testified, "prepared the way for the assault very effectively." Captain Lyman W. V. Kennon of the Sixteenth Infantry claimed the weapons "greatly demoralized the enemy, some of whom could be seen running from their places."[32]

Those artillerymen who denied that the Gatling guns belonged to their arm left themselves with little else to boast about during the Spanish or Philippine wars. The American artillery made a disappointing showing at both El Caney and San Juan. The battery that supported the attack on the first of these two positions wasted hours with an impotent shelling of the

Spanish trenches before turning its attention to the blockhouse that was the key to the position. At San Juan Hill, the charcoal-powder smoke from the American guns gave away their position, and an enemy battery, although weaker, silenced them. Moreover, at least two American company officers reported that these fieldpieces menaced their own infantry as it advanced up San Juan Hill. During the Puerto Rican campaign, rudimentary roads and trails hampered the artillery. While planning an operation on the island in mid-August 1898, Major General James H. Wilson ruled out the utility of either fieldpieces or dynamite guns. After the Philippine War General Frederick Funston recalled the difficulties that the Utah Artillery Battalion had experienced during one of the many small actions of that conflict. Not yet supplied with horses, some of the gunners managed to manhandle two of their fieldpieces into position. Without animals, the struggling crews were limited to dragging their weapons along the middle of an open road, where they soon came under such heavy fire that their officers ordered them to leave the guns temporarily and take cover.[33]

While the Spanish War gave the artillery few opportunities for glory, American military leaders continued to exude confidence in the cavalry. Shortly after the outbreak of the conflict, General Miles proposed to Secretary of War Russell Alger an offensive in Cuba that included a large role for the horse units. Guy V. Henry asserted that Miles recommended sending a full regiment of regular cavalry to Puerto Rico as well, and the veteran yellowleg voiced disappointment that the War Department did not follow the commanding general's advice.[34] Henry, who was among the most rugged troopers of the Indian wars, boasted that a regular cavalry regiment and a Hotchkiss battery could have defeated every Spanish command along the west coast of the island—from Ponce to Arecibo and perhaps to Dorado—in less than two weeks.[35] After the surrender of Santiago, the officers of the Rough Riders determined to continue training for mounted charges against the Spanish, whose soldiers and mounts were smaller than the Americans. "In my regiment the officers began to plan methods of drilling the men on horseback," Theodore Roosevelt recalled, "so as to fit them for use against Havana in December. . . . We were certain that if we ever got a chance to try shock tactics against them they would go down like ninepins, provided only that our men could be trained to charge in any kind of line, and we made up our minds to devote our time to this."[36]

This optimism carried over into the Philippine War, which prompted another round of procavalry sentiment. Indian wars veteran Edward J. McClernand praised a fellow trooper's operations in 1899. "I noted your use of cavalry in the campaign in the Northern part of Luzon," McClernand wrote to S. B. M. Young, "and rejoiced that our arm at last had an opportunity to show its worth under an officer who understands its use." Another grizzled cavalry veteran, Edward S. Godfrey, criticized the army's limited use of horsemen against the Filipinos. "What a great mistake they have made," he contended in a private letter, "in not having more Cavalry over [in the Philippines.] Its efficiency and great value is now clearly seen." Frederick Funston recalled an episode of the Philippine War that supported Godfrey's opinion. When the Americans pursued the Filipinos across a stretch of open ground near Malinta, Funston regretted that his brigade did not have a regiment of cavalry. "Whatever doubt there may be as to the possibility of cavalry charging infantry in line," he reflected, "there is none as to what it can do if it gets in on the flank of a disordered and retreating force. But we had but little cavalry, and that not properly mounted."[37]

The rapid victory over Spain also fueled the nationalistic spirit that had been burning in the United States Army for decades. For the first time in more than eighty years, American soldiers could boast of a triumph over a European opponent, while overlooking the fact that they had defeated a country whose military strength had long been in decline. S. B. M. Young characterized the brief engagement at Las Guasimas as "a test of valor of the opposing forces," which awarded the Americans "the spirit of superiority." A colonel boasted after San Juan Hill that his officers and men had "fought with a steady and determined valor worthy of their country and race." (This commander was perhaps entitled to his chauvinism: his regiment had lost more than a quarter of its numbers in this famous charge.) Another officer, John H. Parker, appealed to nationalistic pride to promote the value of the Gatling gun. He argued that before the Spanish-American War "foreign pen soldiers" had disparaged the machine gun and their criticisms had misled some American officers. The conflict of 1898, Parker proudly insisted, had proven that the rapid-firing weapons were invaluable, that their critics from abroad had been wrong about them, that "the American Regular makes tactics as he needs them," and that the "deductions based on the drill-made automatons of European armies" did not apply to the military forces of the United States.[38]

During the Spanish War American soldiers faced the daunting problem of attacking strong entrenchments. Major General Joseph Wheeler was relieved that the Spanish surrender of Santiago obviated an American assault on the city's defenses. Riding into Santiago for the first time, he was impressed by the strength of their entrenched positions and barricaded roads.[39] "Manila is strongly fortified," warned another American general who was in front of the city in early July 1898. The Spanish prepared elaborate defenses during the Puerto Rican campaign as well. One of Major General James H. Wilson's scouts alerted him that the enemy had mined the route between Aibonito and Coamo and had protected the latter town with sandbags piled eight feet high, ditches dug three and a half feet deep, and iron water pipes pulled across the road. General Wilson also reported: "[Lieutenant Colonel John Biddle] has personally seen four or five points of fortification extending all along the crests from Asomante, well around toward the South and Southeast, and has no doubt we shall find them well prepared in that direction."[40] During the opening months of the Philippine War the Filipinos relied on conventional rather than guerrilla tactics and fought from behind strong entrenchments. Frederick Funston recalled, for example, how the insurgents defended a wagon-road crossing of the Tuliajan River, with fieldworks protecting not only their front but also their flanks.[41] Funston also described an episode in which five American infantry companies came up against a well-entrenched Filipino position. "The enemy in his excellent trenches," he wrote, "was pouring into us a fire that we could not hope to overcome by merely firing back at him."[42]

These turn-of-the-century field fortifications sometimes included barbed wire, an ominous harbinger of its extensive use on the Western Front of World War I. The Spanish employed this entanglement during the Cuban insurrection, before the Americans intervened on the island in

1898. A few days after the Americans had carried San Juan Hill, Brigadier General Jacob F. Kent described the blockhouse on its summit as "a loopholed brick fort surrounded by barbed-wire entanglements."[43] General Wheeler rode into Santiago after its surrender and found at least four places where the Spanish had stretched barbed wire across the road into the city. The sixty-two-year-old Confederate veteran had never seen such obstacles on any Civil War battlefield. "They were not merely single lines of wire," he marveled, "but pieces running perpendicularly, diagonally, horizontally, and in every other direction, resembling nothing so much as a huge thick spider's web with an enormous mass in the center." Brigadier General Thomas M. Anderson anticipated encountering such defenses and, before leaving San Francisco for the Manila campaign, purchased wire cutters with insulated handles for the pioneer parties of his division.[44]

The likelihood that the Spaniards would prepare such strong defenses did not discourage America's military leaders from advocating offensive strategies and tactics. William McKinley, who had been a company officer during the Civil War, in 1898 was commander-in-chief of the country's armed forces. He and his advisers determined that the army should undertake a major offensive, even before most of the troops needed for it had mustered into service. After the war General Miles proudly boasted that the Americans had been "aggressive from start to finish." General Wheeler, a combative Southerner, pushed his own division ahead of Brigadier General Henry W. Lawton's to begin the fire fight at Las Guasimas, perhaps, as at least one historian has suggested, to gain credit for the first action in Cuba. In _The Rough Riders_, Theodore Roosevelt treated offensive tactics as an unquestionable virtue. Reflecting on the skirmish at Las Guasimas, Roosevelt admitted that at first he did not know where the Spanish were, or how the fighting was progressing, but decided that "it could not be wrong to go forward." Writing about the far more famous attack up the San Juan Heights, the future president recalled that again he had already determined on his own to advance, "when Lieutenant Colonel [Joseph H.] Dorst came riding up through the storm of bullets with the welcome command 'to move forward and support the regulars in the assault on the hills in front.'" Early in the Philippine War, Brigadier General Irving Hale advocated the most vigorous of offensive operations, the infantry charge. "Impress upon the men," he advised in March 1899, "that against the present enemy the charge is the safest form of attack."[45]

Just as many officers recognized the dangers of taking the offensive and yet maintained their faith in it, they also made a mixed assessment of the bayonet. On the one hand they acknowledged that it was rarely used; on the other they remained reluctant to discard it. Inspector General Joseph C. Breckinridge questioned many American officers about the bayonet and concluded that soldiers did not use it during the Spanish War, except "as an intrenching tool, or to grind coffee." Breckinridge, however, recommended not abandoning the edged weapon but instead making it "smaller, lighter, and sharper." George R. Fisher, who served with the Utah Light Artillery during the Manila campaign, believed the infantrymen near his battery had confidence in their bayonets, fixing them "for a charge if things came to the worse." An attack by Frederick Funston's Twentieth Kansas on February 7, 1899, at the opening of the Philippine War, illustrated the weapon's uncertain place in the army. The regiment employed it as well as rifle fire in a successful assault against Filipino entrenchments along the Lico Road near Manila, yet Funston confessed that the episode was the only time during his career he saw the bayonet used.[46]

Closing against an opponent with the bayonet, or conducting any offensive operations, depended on locating the enemy's position by effective reconnaissance. During the Spanish War American officers gained useful information from scouting by engineers and other soldiers and from questioning local civilians. Because the conflict proved brief, they did not organize any formal native-scout units. During the longer and bloodier Philippine War the United States Army drew on the precedent of the Indian wars and enlisted the help of Filipino auxiliaries. The first of these units were the Macabebe Scouts, allies of Spain until the end of the Spanish-American War, when they began serving the Americans[47] and quickly proved valuable as guides, scouts, boatmen, and detectives. Building on this precedent, the United States Army created a permanent force of Philippine Scouts,[48] whose numbers grew to 5,000 in about two years. One American veteran of the Philippine War observed after the close of the conflict: "The scouts have become almost an integral part of the regular army."[49]

A number of American officers praised the Philippine Scouts. Lyman W. V. Kennon took pride in being called the "father" of the Ilocano Scouts, recruited largely in northern Nueva Ecija province. "These Ilocanos were splendid marchers and fighters," Frederick Funston claimed, "and were as trustworthy as the Macabebes, and that is saying a great deal." After the Philippine War moved from conventional conflict to guerrilla warfare, the Americans needed help identifying their enemies. Officers like Brigadier General Frederick Dent Grant, son of the famous Civil War commander, concluded that the auxiliaries proved valuable in this new role. "The native population was . . . taught," Grant reported, "that in the Macabebe scouts the United States has a loyal servant who can be depended on to pick out of a crowd of natives, however large, all the insurgents masquerading as 'amigos' and the culprits from other provinces." After the Philippine War ended, the veteran William P. Duvall regreted the "simple truth . . . that the Scouts are, to put it mildly, generally not appreciated and not understood."[50]

Although the Philippine Scouts helped find the enemy, they could not resolve the difficulties of carrying his positions. The Americans still faced all the dangers of frontal assaults against well-armed defenders, and they looked anxiously for alternatives. Major General Nelson A. Miles, who had survived two of the Civil War's bloodiest headlong attacks, at Fredericksburg, where he had been wounded, and at Cold Harbor, was determined to avoid such tactics when he directed the Puerto Rican campaign. Throughout this operation, Miles strongly emphasized flanking movements rather than direct assaults.[51] "Envelop or outflank the enemy rather than attack in front," Miles enjoined his subordinate John R. Brooke, "and under no circumstances assault entrenched lines." Brooke had also fought at both Fredericksburg and Cold Harbor, where he had been severely wounded, and his cautious conduct in Puerto Rico suggests that he readily understood his superior's counsel. Brooke spent several days preparing to advance against an entrenched Spanish position on the road between Guayama and Cayey, so long that the news of the peace protocol obviated any attack.[52] Theodore Roosevelt, who as much as any American soldier favored aggressive operations, hoped that the Americans could take Santiago by siege rather than by frontal attacks. "If the city could be taken without direct assault on the entrenchments and wire entanglements," he acknowledged, "we earnestly hoped it would be, for such an assault meant, as we knew by past experience, the loss of a quarter of the attacking regiments." General Wesley Merritt similarly wanted to avoid an attack against the defenses of Manila, unless he got support from the navy. He concluded that unless the heavy guns on ships could breach the works and demoralize the defenders, an infantry advance was too dangerous.[53]

When the Americans *did* make frontal assaults, they often moved in rushes, rather than attempting any sustained advance. The commander of the First California described how his unit made use of a loose formation and rush tactics during the early operations against Manila. "The battalion moved in column of fours along the Calle Real to the Parsay road," Colonel James S. Smith reported, "and were then thrown out in extended order between the Calle Real and the beach. Within 100 yards of this point the battalion advanced by rushes, under a galling fire by the enemy, to the old intrenchments previously occupied by the [Filipino] insurgents." Early in the Philippine War, Brigadier General Irving Hale advised his men: "When under fire, advance by alternate rushes of companies or platoons, thus keeping the enemy down by a practically continuous fire, to proper distance and then charge." Frederick Funston practiced the advance-by-rushes tactic time and again, during operations against the Filipinos. Relating combat to training, Funston recalled: "We put into practice what we had learned on the drill ground at the Presidio. One platoon, that is half a company, would rush forward for about fifty yards and throw itself prone, while the other platoon would rise and rush past it." He recalled another episode, when the combination of a double-time march and the advance-by-rushes exhausted his Twentieth Kansas so badly that dozens of soldiers dropped out of the ranks. This tough-minded combat officer also described yet another instance when his regiment began an attack, "advancing by rushes in the orthodox manner."[54]

Attackers had to contend against the deadly firepower of late nine-teenth- century weapons, the difficulties of terrain, and the stress of combat, which meant that assaults were often made without much regard for formal order or tactics. The most famous attack of the war, the charge up San Juan Heights, was an irregular effort. The attackers, wrote Richard Harding

Davis, "had no glittering bayonets, they were not massed in regular array. There were a few men in advance, bunched together, and creeping up a steep, sunny hill, the tops of which roared and flashed with flame." Farther behind these first few troops, "spreading out like a fan, were single lines of men, slipping and scrambling in the smooth grass, moving forward with difficulty, as though they were wading waist high through water, moving slowly, carefully, with strenuous effort."[55] Davis contrasted the actual event with the scene depicted in so many "picture-papers," with the Americans "in regular formation, rank after rank."[56] "The final advance up the spur [of San Juan Hill]," Captain Lyman W. V. Kennon reported, "was without any regular formation, nor was any attempted; it was a paramount consideration to get men on the summit, with or without regular formation." Private Post agreed that there "was no nice order, no neatly formed companies crossing that plain or mounting the slope." A man in the ranks, Post offered a commonplace analogy: "It was more like a football field when the game is over and a mess of people are straggling across it, except that these men were on the run, yelling, and with no time to lose."[57]

American soldiers recognized that attackers as well as defenders could take advantage of the enormous firepower infantry could now deliver. During both the Spanish and Philippine wars, assault troops tried the innovation of firing while advancing. One captain reported after the Santiago campaign that his men began firing as they went up San Juan Hill and admitted that he could not stop them. Frederick Funston decided, early in the Philippine War, that this tactic had merit. During their February 7, 1899, attack, his soldiers had discouraged their opponents with firepower before closing with the bayonet. "Our men . . . ," Funston recalled "fairly combed the top of that dike with bullets. We were advancing at a walk and it was point-blank range, and our fire so disconcerted the enemy that though they plied their rifles with great vigor, they were not exposing themselves enough to get any sort of good aim."[58] Ten days later at Caloocan, the hard-fighting Kansan directed his regiment to make a sustained assault, relying on continuous firepower, rather than advancing in rushes and firing in volleys. "The experience obtained in our attacks of the 5th and 7th [February, 1899]," Funston later wrote, "had convinced me that by sweeping the ground that we were advancing over with a storm of bullets we could so demoralize the enemy that his fire would be badly directed." Advancing part of the time "at a walk, the men firing to the front as rapidly as they could," the Kansans carried the Filipino entrenchments.[59] He also pointed to another attack in which his men suppressed an entrenched defense with heavy firepower. "The Filipinos were kept down in their trenches by the fire poured in upon them," Funston wrote of this episode, "so that they simply could not rise up to take any aim at all." The pragmatic Kansan on some occasions used the advance by rushes and volleying and on others the steady approach and sustained fire. He came to believe that the latter method held at least one advantage: "The shooting is much more accurate."[60]

Funston's use of the continuous advance accompanied by heavy firepower anticipated by four decades the description given by famous World War II commander, Lieutenant General George S. Patton, Jr., of "marching fire." Funston's experience at the turn of the century presaged General Patton's ideas about how infantry should attack and about how defenders would be affected. "The proper way to advance," Patton declared, "particularly for troops armed with that magnificent weapon, the M-1 rifle, is to utilize marching fire and keep moving. . . . The whistle of the bullets, the scream of the ricochet, and the dust, twigs, and branches which are knocked from the ground and the trees have such an effect on the enemy that his small-arms fire becomes negligible."[61]

Despite the army's stress on marksmanship and individual fire during the peacetime decade of the 1880s, American leaders relied on volley fire during the Spanish and Philippine wars. Many commanders feared that if their men, who were now armed with repeaters, were allowed to fire at will, their officers would lose control over them and the troops would fire recklessly and waste ammunition. Brigadier General Irving Hale followed this thinking in a general order that he issued, early in the Philippine War. "Controlled volley firing by Company, platoon, or squad will be used," Hale directed, "except when the enemy is scattered and retreating, and individual firing, when necessary[,] will be closely regulated." An officer of the Seventh Infantry reported that at El Caney the regiment delivered precise volleys, while acknowledging that some companies resorted to firing by two-man files, but only after they had run low on ammunition. Captain

Lyman W. V. Kennon, who had been interested in tactical theory before the Spanish-American War, commanded a company of the heavily engaged Sixteenth Infantry during the attack on the San Juan Heights and prepared a thoughtful report on the battle. Kennon judiciously concluded: "As to results, it is difficult to give a decided opinion, though the effect of volley firing seemed to be somewhat greater than that of individual firing."[62]

While they continued to rely on volley fire, officers looked to new technologies to help attackers overcome the advantages of defenders. Some American soldiers hoped that the Sims-Dudley dynamite gun that accompanied General Shafter's corps would blast apart the defenses of Santiago. Colonel Leonard Wood spoke favorably of this weapon, and its commanding sergeant claimed for it "the destruction of three Spanish guns, the extensive demolition of trenches, and presumably a considerable loss of life to the enemy." The dynamite gun, however, remained an experimental weapon and it proved difficult to evaluate its effectiveness. The Sims-Dudley fieldpiece fired only twenty rounds during the entire Santiago campaign. While its commander claimed its field test a success, he also acknowledged, and candidly described to his superiors, its "faults in material and construction."[63]

Other inventions also offered to help attackers gain parity with defenders. Some commanders took interest in the field telephone, but the instrument's technology had not matured enough to make it as helpful in 1898 as it would prove during the twentieth century. Signal Corps officers established a telephone line for General Shafter during the Santiago campaign, allowing the Fifth Corps commander to talk to his immediate subordinates. This communication system, however, did not extend to the brigade commanders and the officers below them in the chain of command, the leaders who in fact exercised the greatest control over the San Juan and El Caney battles.[64] Similar arrangements prevailed during the operations against Manila. The chief signal officer of the Eighth Corps detailed men to the division leader Brigadier General Thomas M. Anderson, his two immediate subordinates, and his reserve commander and dispatched troops to a point on the beach where they could communicate with the American ships.[65] Again as in Cuba, there was no communication with small units,[66] and furthermore, disruptions of the lines, shortages of insulators and wires, difficult terrain, bad weather, interference by Filipino revolutionaries, and other problems hindered the system.[67]

The military balloon, the forerunner of twentieth-century air power that had been introduced by "Professor" T. S. C. Lowe during the Civil War, represented another emerging technology that in 1898 could offer only limited help to attacking infantrymen. During the Santiago campaign, Lieutenant Colonel Joseph E. Maxfield and a twenty-four-man balloon detachment accompanied the Fifth Corps as it approached the San Juan Heights. These airborne observers provided their comrades on the ground with two useful pieces of intelligence: a confirmation that the Spanish held San Juan Hill with a sizable force and the identification of a trail that offered the American infantry an alternative attack route.[68]

Unfortunately for the Americans, the balloon also proved as much hindrance as help, after Lieutenant Colonel George F. Derby, Maxfield's superior, ordered it towed too close to the front. Obeying this unfortunate instruction, the detachment temporarily disrupted the advance of the First and Tenth Cavalry and found itself in terrain where it could not raise the balloon without entangling its guide ropes in branches and undergrowth. The craft became immobilized and the Spanish soon punctured it with rifle and artillery fire. A veteran of the Sixth Cavalry estimated that the huge bag was not aloft more than fifteen minutes before the enemy shot it to pieces. Worse yet for the Fifth Corps, the balloon's approach and its last position identified the American infantry's line of advance and the Spanish quickly used it as an aiming point. "The trail rope led directly down into the Aguadores road . . . ," wrote Private Post. "It was a beautiful range marker for the Spanish artillery and infantry, and they promptly used it as such. . . . The balloon rope had given the Spanish all the information they wanted, and they . . . concentrated all their fire power on that trail."[69]

The use of the balloon and field telephone during the Santiago campaign pointed to technologies that would help attackers in the future, but events on Spanish-American War battlefields showed that defenders still held commanding advantages. During the Cuban campaign the Americans eventually carried El Caney, but they needed overwhelming numbers and ten hours to do so. Brigadier General Joaquín Vara del Rey held the position with three companies of Spanish regulars and one of guerrillas, only about 500 defenders in all, unsupported by artillery,[70] while Brigadier General Henry W. Lawton brought a division of 5,400 against him.[71] Lawton promised General Shafter he would take El Caney in two hours, but the defense held for more than ten, from 6:15 in the morning until about 4:30 in the afternoon.[72]

The Americans also carried the San Juan Heights, in the most famous attack of the war, but General Linares's deployment of his forces contributed much to their success. The Spanish troop dispositions suggested that their commander had in mind little more than a delaying action. Linares's chief intentions in stationing his men were to cooperate with his comrades Admiral Pascual Cervera, whose fleet had taken refuge in Santiago's harbor, and with Colonel Frederico Escario, who was marching 3,500 soldiers to reinforce Linares, and also to contend against the force of 5,000 Cuban insurgents in the area. The Spanish commander gave more weight to these considerations than to concentrating his forces on the San Juan Heights. Linares probably hoped not to defeat the American army in a pitched battle but rather to inflict casualties on his opponent, buy time while Cervera decided what to do, stall his approaching foe in front of Santiago, and link his own forces with Escario's column.[73]

The San Juan position was an outpost and not the main line of Santiago's defenses. This fact helps explain the success of the American attack. It sheds light on the Spanish decision to entrench the actual, instead of the military, crest of the high ground. Preparing for no more than a hasty delaying action, General Linares's men fortified the summit of San Juan Hill, rather than the highest point on the slope from which defenders would have a clear view to fire down on their assailants. When the attacking Americans were "at the foot of [San Juan] Hill, the Spaniards were unable to see them," Private Post wrote, "and from the base to the military crest ... were as safe as if they had been back in Siboney. From the military crest to the actual crest was but twenty to thirty feet and no charge can be stopped within such a distance." Captain Lyman W. V. Kennon, a company commander, agreed with Post: "The enemy's trenches were not on the 'military crest' of the hill, but farther back. A very considerable dead space was thus left unexposed to fire."[74] Another captain reported that he directed the soldiers around him into this same "dead space at the foot of the hill."[75]

A view along the trenches during the Santiago campaign, 1898. The Spaniards built similar fieldworks to protect the city, and Theodore Roosevelt asserted that the Americans would suffer terribly if they attacked against the defenders' rifle fire. (Photograph courtesy of United States Army Military History Institute)

The attackers succeeded, but the cost was high. One thorough historian of the conflict has estimated the American losses at the San Juan Heights and El Caney at more than 220 officers and men killed and 1,300 wounded. Of General Wheeler's division 6 officers were killed and 26 wounded,[76] and the Tenth Cavalry suffered total casualties of 80 officers and men. General Jacob F. Kent used a phrase familiar to prewar theorists when he reported that his division passed "through a zone of [the] most destructive fire" in front of the San Juan Heights. One of his regiments, the Sixth Infantry, lost nearly a quarter of its numbers within about ten minutes, and its neighbor the Sixteenth Infantry suffered an even higher percentage. The adjutant of the Sixteenth made a grim entry in his diary shortly after the assault: "Capt [Thomas C.] Woodbury was shot in leg. Capt [Theophilus W.] Morrison was killed. [First Lieutenant Robert E. L.] Spence shot in arm[.] [First Lieutenant] Dennis Michie wounded through bowels[.] The regiment had a very hot & hard day of it."[77]

After the San Juan Heights fighting, Shafter and his soldiers lost some of their enthusiasm for combat. The attackers had suffered so heavily that the army commander and his subordinates, rather than undertaking a siege of Santiago with confidence, were demoralized enough to consider a retreat. Roosevelt, whose Rough Riders had lost eighty-seven killed and wounded, understood clearly the cause of their despair. After the Fifth Corps assaults, he wrote to Henry Cabot Lodge: "the Spaniards fight very hard and charging these entrenchments against modern rifles is terrible."[78] Another American officer described the sordid aftermath of the struggle for the San Juan Heights: "The stench along all [the] roads from dead or wounded men or animals and general garbage and decaying matter was sickening, which, together with bad water, commenced to tell on the men."[79]

Before the Spanish War, the desirability of attacking entrenchments had been an important question of tactical theory. For the soldiers who survived such combat, attacks on entrenchments were no longer an academic subject but a traumatic reality. An American general whose brigade advanced against El Caney wrote, shortly after the assault: "The attack of a fortified place by Infantry is usually attended by disaster, and is recognized as one of the most difficult military operations." Reflecting on his experience in front of San Juan Heights, a veteran of the Sixth Cavalry cut through the theory to the reality: "Lord knows how many shoot at you." During the charge, the same soldier recalled, he had "felt as big as a barn." One survivor of the San Juan action vehemently criticized his commander's frontal-assault tactics. "Gen Shafter is a fool," wrote an officer of the Sixteenth Infantry, "and I believe should be shot."[80]

Lieutenant John H. Parker reflected that before the Spanish-American War, the army had engaged in "[v]ery able discussions" of "the theoretical changes of the battlefield" brought by improved weapons and entrenchments, "but no proper conclusion had been reached."[81] His statement remained equally true after the brief conflict of 1898. The Spanish War convinced all but the most reactionary American officers of the perils of "crossing the deadly ground" in front of entrenched defenders, yet the tactical theorists could not reach any "proper conclusion" about coping with the increasing complexities of warfare.

This challenge became more formidable as the United States Army entered the twentieth century. By 1898 the blue-clad, Indian-fighting constabulary had faded into the past, and the American military faced the dangers of a new era. After the Spanish War, some of the army remained in the Philippines, at first conducting conventional operations and later fighting a guerrilla war, a harbinger of a conflict decades later in Southeast Asia. Fifteen years after the Philippine War, the United States would enter World War I. American soldiers would suffer the horrors of the Western Front, where trenches, barbed wire, rifles, machine guns, and artillery would make crossing the "deadly ground" more harrowing than ever in human history.

Ignorant of this future, the United States Army of 1898 enjoyed its triumph over Spain. While the troop ships ferried the young American soldiers back from Cuba to Long Island, far across the ocean in the quiet farmlands of northern France, the birds moved among the small forests near Cantigny and through the trees of Belleau Wood.

Notes

Preface

1. Robert Home, *Précis of Modern Tactics* (London, 1882), 70–71.

1. *No More Cold Harbors*

1. The chaplain of the First Massachusetts is quoted by Edward Steere, *The Wilderness Campaign* (Harrisburg, Pa., 1960), 226, citing W. H. Cudworth, *History of the First Massachusetts Infantry* (Boston, 1886), 460. On the dominance of the tactical defensive during the Civil War, see Grady McWhiney and Perry D. Jamieson, *Attack and Die: Civil War Military Tactics and the Southern Heritage* (University, Ala., 1982), and Edward Hagerman, *The American Civil War and the Origins of Modern Warfare: Ideas, Organization, and Field Command* (Bloomington and Indianapolis, 1988). For Burnside's and Grant's losses, see Thomas L. Livermore, *Numbers and Losses in the Civil War in America, 1861–65* (Boston and New York, 1900), 96, 114.

2. On the resilience of Civil War armies, see Herman

Hattaway and Archer Jones, *How the North Won: A Military History of the Civil War* (Urbana, Ill., 1983), 47, 168, 200, 229, 384, 454, 692, 720. Hood's losses at Franklin are discussed in James Lee McDonough and Thomas L. Connelly, *Five Tragic Hours: The Battle of Franklin* (Knoxville, 1983), 157. On the importance of the breech-loading rifle, see Perry D. Jamieson, "The Development of Civil War Tactics" (Ph.D. diss., Wayne State University, 1979), 186, 188, 193–195; on the diminished value of the bayonet and saber, 186, 213–214. Although American soldiers readily understood the advantages of the breechloader, the U.S. Army was slow to adopt a repeating breechloader as its standard arm. See Richard I. Wolf, "Arms and Innovation: The United States Army and the Repeating Rifle" (Ph.D. diss., Boston University, 1981).

3. *Army and Navy Journal* (hereinafter *ANJ*) 3 (November 4, 1865), 169; "Veteran," "Change of Tactics," *ANJ* 3 (September 23, 1865), 76; Ulysses S. Grant to Edwin M. Stanton, February 4, 1867, Roll 680, Grant Board Papers, Adjutant Generals Office (AGO) File 312 A 1869, M-619, National Archives; John Pope, Address to the Army of the Tennessee, October 16, 1873, Roll 19, General Correspondence, William T. Sherman Papers, Library of Congress.

4. William T. Sherman to Stephen A. Hurlbut, May 26, 1874, Roll 46, Letterbooks, Sherman Papers; Annual Reports of the War Department, 1822–1907, M-997, National Archives, Report of the Artillery School, September 12, 1871, Roll 18, 1:79–80 (hereinafter AR, used to mean vol. 1 only); William T. Sherman to George W. Getty, January 10, 1878, Roll 45, Letterbooks, Sherman Papers; AR, Report of the Artillery School, October 18, 1879, Roll 33, 180–181, (quotation, 180). Founded in the 1820s and active until 1860, the Artillery School was revived in 1868. Russell F. Weigley, *History of the United States Army* (Bloomington, Ind., 1984), 273.

5. Philip H. Sheridan to William T. Sherman, August n.d. and August 20, 1870, Roll 15, Correspondence, Sherman Papers; Philip H. Sheridan, *Personal Memoirs of P. H. Sheridan*, 2 vols. (New York, 1888), 2:451. See also Philip H. Sheridan to William T. Sherman, December 26, 1870, Roll 16, Correspondence, Sherman Papers.

6. Weigley, *History of the Army,* 290; *ANJ* 3 (November 4, 1865), 169; Emory Upton, *A New System of Infantry Tactics, Double and Single Rank, Adapted to American Topography and Improved Fire-Arms* (New York, 1867).

7. *ANJ* 3 (November 4, 1865), 169; "Veteran," "Change of Tactics," 76; McWhiney and Jamieson, *Attack and Die,* 81–88; H. S. Hawkins, quoting Colonel Robert Home, in "Outline of a Manual of Infantry Drill," *Journal of the Military Service Institution of the United States* (hereinafter *JMSIUS*), 11 (1890), 361.

8. Emory Upton to Dear Brother, November 6, 1863, and to My Dear Sister, June 4, 1864, and June 5, 1864, in Peter S. Michie, *The Life and Letters of Emory Upton* (New York, 1885), 80, 108, and 109.

9. Michie, *Life and Letters of Upton*, 189–190; U.S. War Department, *The War of the Rebellion: A Compilation of the Official Records of the Union and Confederate Armies*, 128 vols. (Washington, D.C., 1880–1901), ser. 1, 29, pt. 1:576, 586, 588, 589; 36, pt. 1:667–668.

10. Michie, *Life and Letters of Upton*, 156, 191; Emory Upton to Edward D. Townsend, January 13, 1866, ibid., 191–193. On Upton's ideas about tactics, see also Stephen E. Ambrose, *Upton and the Army* (Baton Rouge, 1964), 22, 60–64.

11. "Upton's Tactics," *ANJ* 4 (September 29, 1866), 85; Henry B. Clitz to Edward D. Townsend, January 14, 1867, Roll 680, Clitz Board Papers, AGO File 312 A 1869, M-619, National Archives. Before it considered Upton's tactics, the Clitz Board had reviewed and rejected a system proposed by William H. Morris. Henry B. Clitz to Edward D. Townsend, January [n.d.] 1867, ibid.

12. Special Orders No. 300, June 11, 1867, Edward D. Townsend to Edwin M. Stanton, February 6, 1867, Ulysses S. Grant to Edwin M. Stanton, February 4, 1867, and Proceedings of the Grant Board, Roll 680, Grant Board Papers; Michie, *Life and Letters of Upton*, 197, 198; Upton, *A New System*, ii, iii–iv; William T. Sherman to Philip St. George Cooke, March 7, 1876, Roll 46, Letterbooks, Sherman Papers.

13. AR, Report of the General of the Army, November 10, 1870, Roll 17, 5–6; Special Orders No. 60, August 6, 1869, Roll 682, Schofield Board Papers, AGO File 312 A 1869, M-619, National Archives; Russell F. Weigley, "Emory Upton," *Dictionary of American Military Biography* (Westport, Conn., 1984), 3:1123–1124; Emory Upton to John M. Schofield, September 13, 1869, Roll 682, and William T. Sherman to William W. Belknap, January 18, 1871, Roll 685, Schofield Board Papers; "New Cavalry Tactics," *ANJ* 11 (June 27, 1874), 730; Emory Upton to Thomas M. Vincent, July 11, 1873, Roll 685, Schofield Board Papers.

14. "Where Are the Tactics?," *ANJ* 10 (September 7, 1872), 55; "New Cavalry Tactics," 730.

15. William T. Sherman to Emory Upton, January 3, August 18, and September 23, 1873, Roll 45, and May 21, 1873, Roll 46, Letterbooks, Sherman Papers.

16. Emory Upton, *Infantry Tactics, Double and Single Rank* (New York, 1874); U.S. War Department, *Cavalry Tactics, United States Army, Assimilated to the Tactics of Infantry and Artillery* (New York, 1874), and idem, *Artillery Tactics, United States Army, Assimilated to the Tactics of Infantry and Cavalry* (New York, 1874);

William T. Sherman to William H. Morris, August 17, 1882, Roll 47, Letterbooks, Sherman Papers. On the "assimilated" tactics, see also Ambrose, *Upton and the Army*, 76–81.

17. Upton, *A New System*, 1, 48–49, 57–59, 59–63, 83–87. Upton's first manual and his later "assimilated" one both assumed a regiment of ten companies. Ibid., 1–2, and Upton, *Infantry Tactics*, 149.

18. Upton, *A New System*, 92–96; Emory Upton to Edward D. Townsend, January 13, 1866, in Michie, *Life and Letters of Emory Upton*, 192; "Line Officer," "A Tactical Necessity," *ANJ* 23 (January 16, 1886), 488.

19. Upton, *A New System*, iii, iv; "Upton's Tactics," *ANJ* 4 (September 29, 1866), 85, and (February 2, 1867), 421; "Advantages of Upton's Tactics," *ANJ* 5 (May 23, 1868), 634.

20. William H. Morris to Edwin M. Stanton, January 4, 18[6]7, Roll 680, Clitz Board Papers; T. W. Sherman to Lorenzo Thomas, March 26, 1867, Roll 680, Grant Board Papers; "Atlanta," "Upton's and Casey's Tactics," *ANJ* 5 (June 27, 1868), 714. On Upton's critics, see also Jamieson, "Development of Civil War Tactics," 204–207.

21. George B. McClellan, "Army Organization," *Harper's New Monthly Magazine* 49 (1874):409.

22. "Bayonet and Breech-loader," *ANJ* 6 (November 14, 1868), 200; William T. Sherman to Philip H. Sheridan, February 20, 1878, and J. W. Reilly to Philip H. Sheridan, April 2, 1878, Roll 45, Letterbooks, Sherman Papers.

23. Francis J. Lippitt, *A Treatise on the Tactical Use of the Three Arms: Infantry, Artillery, and Cavalry* (New York, 1865), 5, 24 for quotations, and see also 4, 8, 27, 51; Board of Officers, *Reports of Experiments with Rice's Trowel Bayonet, Made by Officers of the Army, Pursuant to Instructions from the War Department* (Springfield, Mass., 1874), 17; William T. Sherman to Francis V. Greene, October 20, 1879, Roll 45, Letterbooks, Sherman Papers.

24. Board of Officers, *Rice's Trowel Bayonet*, 34, 39.

25. McWhiney and Jamieson, *Attack and Die*, 131–132; "Rifle and Sabre," *ANJ* 6 (October 31, 1868), 131; William T. Sherman to Philip H. Sheridan, February 20, 1878, Roll 46, Letterbooks, Sherman Papers.

26. John M. Schofield to William T. Sherman, October 12, 1869, Roll 15, Correspondence, Letterbooks, Sherman Papers; Ordnance Memoranda No. 11, June 10, 1870, Roll 685, Schofield Board Papers; "Small Arms and Accoutrements," *ANJ* 8 (June 17, 1871), 700; Report of Captain E. C. Hentig, August 14, 1880, Roll 94, Recommendations of Lt. Col. James W. Forsyth, AGO File 1558 AGO 1882, M-689, National Archives.

27. William R. Parnell, "Sabre Practices—Cavalry Tactics—1841," *ANJ* 8 (April 22, 1871), 571; "Caballo," "Rifle and Sabre," *ANJ* 6 (January 9, 1869), 326; "A Volunteer Cavalryman," "The St. Louis Board," *ANJ* 8 (July 1, 1871), 735; "Sabre of the Regulars," "Sabres for the Cavalry," *ANJ* 14 (October 14, 1876), 154. See also C. C. C. Carr, "Discussion," *Journal of the United States Cavalry Association* (hereinafter *JUSCA*) 1 (1888), 53–54.

28. Wesley Merritt, "Our Cavalry," *ANJ* 16 (July 5, 1879), 873; John C. Kelton to the Adjutant General, May 24, 1880, Roll 94, Recommendations of Lt. Col. James W. Forsyth.

29. McWhiney and Jamieson, *Attack and Die*, 130; "Sabre of the Regulars," "Sabres for the Cavalry," 154; "A Volunteer Cavalryman," "The Lessons of the Decade," *ANJ* 8 (January 21, 1871), 366.

30. McWhiney and Jamieson, *Attack and Die*, 135–136; James H. Wilson to William H. Emory, July 27, 1868, Roll 680, Emory Board Papers, AGO File 312 A 1869, M-619, National Archives; William T. Sherman to Philip H. Sheridan, February 20, 1878, Roll 46, Letterbooks, Sherman Papers.

31. Francis B. Heitman, *Historical Register and Dictionary of the United States Army*, 2 vols. (Washington, D.C., 1903), 1:763; Report of First Lieutenant Gilbert E. Overton, May 31, 1880, Roll 94, Recommendations of Lt. Col. James W. Forsyth; *ANJ* 13 (October 16, 1875), 153; John B. Hood, *Advance and Retreat* (Bloomington, Ind., 1959), 132.

32. Jamieson, "Development of Civil War Tactics," 218; U.S. War Department, *Cavalry Drill Regulations, United States Army* (Washington, D.C., 1891), 10.

33. "American Infantry Tactics," *ANJ* 3 (October 28, 1865), 149; "A Few Thoughts on Artillery," *ANJ* 8 (July 15, 1871), 768; "The Modern Breechloader," *ANJ* 15 (April 13, 1878), 571. See also "Modern Infantry Fire," *ANJ* 17 (January 3, 1880), 427.

34. McClellan, "Army Organization," 409, 410; Francis J. Lippitt, *A Treatise on Intrenchments* (New York, 1866), 109, 126, 127, 135, 136; "Rank Formation with Breech-Loaders," *ANJ* 4 (February 23, 1867), 421.

35. General Order No. 7, Headquarters Artillery School, March 25, 1878, quoted in AR, Report of the Commanding Officer of the Artillery School, November 4, 1878, Roll 30, 200; David A. Armstrong, *Bullets and Bureaucrats: The Machine Gun and the United States Army, 1861–1916* (Westport, Conn., and London, 1982), 14–22, 60–61.

36. "Tactics for Field Artillery," 11, Roll 682, Proceedings and Report of the Barry Board, AGO File 312 A 1869, M-619, National Archives; "The Gatling Gun," *ANJ* 8 (July 8, 1871), 752–753; Armstrong, *Bullets and Bureaucrats*, 32, 36.

37. Robert V. Bruce, *Lincoln and the Tools of War* (Indianapolis, 1956), 290–291; Armstrong, *Bullets and Bureaucrats*, 18–19.

38. Armstrong, *Bullets and Bureaucrats*, 17–18.

39. Ibid., 14, 15–19, 24–25, 33; Bruce, *Lincoln and the Tools of War*, 120, 196–197, 200, 249–251, 290.

40. McClellan, "Army Organization," 409; "A Volunteer Cavalryman," "The Lessons of the Decade," *ANJ* 8 (April 15, 1871), 558; "Tactics for Field Artillery," 167–170, Roll 682, Proceedings and Report of the Barry Board; Emory Upton to Henry A. DuPont, February 25, 1875, in Michie, *Life and Letters of Emory Upton*, 212; War Department, *Artillery Tactics Assimilated*, 74–79.

41. See David A. Armstrong's thorough study, *Bullets and Bureaucrats*.

42. Robert M. Utley, *Frontier Regulars: The United States Army and the Indian, 1866–1891* (New York and London, 1973), 259, 265; Edgar I. Stewart, *Custer's Luck* (Norman, Okla., 1955), 178, 246; James S. Brisbin to E. S. Godfrey, January 1, 1892, quoted in E. A. Brininstool, *Troopers with Custer: Historic Incidents of the Little Big Horn* (London and Lincoln, Nebr., 1989), 279–280; James H. Bradley, *The March of the Montana Column: A Prelude to the Custer Disaster,* ed. Edgar I. Stewart (Norman, Okla., 1961), 150; Edward J. McClernand, *With the Indian and the Buffalo in Montana, 1870–1878* (Glendale, Calif., 1969), 49 and 50. McClernand believed Custer "could have taken the [Gatling] guns as easily as [John] Gibbon [and Alfred Terry], for the latter crossed a more difficult country." Ibid., 47. Whether the Gatlings would have saved Custer is another matter. See Armstrong, *Bullets and Bureaucrats*, 82.

43. Nelson A. Miles to William T. Sherman, January 4, 1877, Roll 23, General Correspondence, Sherman Papers; William T. Sherman, Address to the Class of 1880, U.S. Artillery School, April 28, 1880, Roll 45, Letterbooks, Sherman Papers.

44. Robert Wooster, *The Military and United States Indian Policy, 1865–1903* (London and New Haven, Conn., 1988), 57–58; William T. Sherman to Emory Upton, January 3, 1873, Roll 44, Letterbooks, Sherman Papers.

45. Wooster, *The Military and Indian Policy*, 57; William T. Sherman to Emory Upton, January 3, 1873, Roll 44, Letterbooks, Sherman Papers; AR, Orders No. 127, Post of Fort Leavenworth, Kansas, May 28, 1884, Roll 47, 192; Edward Bruce Hamley, *The Operations of War* (Edinburgh and London, 1866), 41–45, especially 44–45.

46. Ibid., 44; William T. Sherman to Francis V. Greene, October 20, 1879, Roll 45, Letterbooks, Sherman Papers. It should be acknowledged that while American soldiers were unable to solve the problem of the dominance of the de-

fensive, neither were European military men, many of whom failed to grasp the tactical lessons of the Civil War. See Jay Luvaas, *The Military Legacy of the Civil War: The European Inheritance* (Chicago, 1959), especially 46, 49, 73–74, 115, 123, 140–142, 150, 166–168, 179.

47. Alexander S. Webb, "Through the Wilderness," Robert U. Johnson and Clarence C. Buel, eds., *Battles and Leaders of the Civil War,* 4 vols. (New York, 1956), 4:152; Upton, *Infantry Tactics*, viii.

48. "Uniform Tactics," *ANJ* 7 (September 18, 1869), 66; William H. Morris to George H. Thomas, June 8, 1866, Roll 680, Clitz Board Papers; T. C. H. Smith to Philip St. George Cooke, November 4, 1866, Roll 680, Grant Board Papers.

49. Charles D. Parkhurst, "Field Artillery: Its Organization and Its Role," *Journal of the United States Artillery* (hereinafter *JUSA*) 1 (1892), 262; Jacob D. Cox, *Military Reminiscences of the Civil War*, 2 vols. (New York, 1900), 1:175; William T. Sherman to the Secretary of War, August 20, 1881, Roll 47, Letterbooks, Sherman Papers; H., "Ignorance of Tactics in the Army," *ANJ* 22 (January 3, 1885), 451. This opinionated letter sparked a lively debate in the *Army and Navy Journal*. See: G. N. Whistler, "Ignorance of Tactics in the Army," *ANJ* 22 (January 17, 1885), 480; "Ignorance of the Tactics," *ANJ* 22 (January 24, 1885), 515; and "Knowledge of Tactics," *ANJ* 22 (February 21, 1885), 597.

50. AR, Report of the Department of Missouri, September 15, 1877, Roll 28, 65; Cox, *Military Reminiscences*, 1:177; "Cavalry," "Cavalry Tactics," *ANJ* 10 (February 8, 1873), 410; AR, Report of the Inspector General, October 9, 1880, Roll 35, 50.

51. M. C. Meigs to William T. Sherman, July 9, 1879, Roll 26, General Correspondence, Sherman Papers; AR, Report of the General of the Army, November 1, 1879, Roll 33, 14; AR, Report of the Acting Inspector General, Department of Dakota, October 3, 1879, 67; AR, Report of the Department of the Missouri, October 3, 1879, 84. On the army's experience with Reconstruction, see James E. Sefton, *The United States Army and Reconstruction* (Baton Rouge, 1967), and Edward M. Coffman, *The Old Army: A Portrait of the American Army in Peacetime, 1784–1898* (New York and Oxford, 1986), 234–246.

52. The post–Civil War army's lack of "doctrine," in the formal way that twentieth-century officers defined the term, is discussed in the third chapter of this work and in Larry D. Roberts, U.S. Army Combat Studies Institute, "Strategy, Doctrine, and the Frontier Army" (MS in possession of P. D. Jamieson).

53. William T. Sherman to Philip St. George Cooke, May 10, 1881, Roll 46, Letterbooks, Sherman Papers.

2. *Hard and Dangerous Service*

1. William H. Carter, *From Yorktown to Santiago with the Sixth U.S. Cavalry* (Baltimore, 1900), 279. This point is well articulated by Robert Wooster, *The Military and United States Indian Policy, 1865–1903* (London and New Haven, Conn., 1988), 214.

2. Robert M. Utley, *Frontier Regulars: The United States Army and the Indian, 1866–1891* (New York and London, 1973), 46, 410.

3. Ibid., 236; Don Rickey, Jr., *Forty Miles a Day on Beans and Hay: The Enlisted Soldier Fighting the Indian Wars* (Norman, Okla., 1963), 272.

4. Stephen C. Mills to the Regimental Adjutant, 12th U.S. Infantry, July 12, 1907, Stephen C. Mills Papers, U.S. Army Military History Institute; AR, Report of the Military Division of the Missouri, November 1, 1868, Roll 16, 4; AR, Report of the General of the Army, November 20, 1869, Roll 17, 24; Philip H. Sheridan to William T. Sherman, February 10, 1877, Roll 23, General Correspondence, William T. Sherman Papers, Library of Congress; William B. Hazen to William T. Sherman, May 5, 1877, Roll 24, ibid.; AR, Report of the Department of the Platte, September 30, 1880, Roll 35, 80. See also William T. Sherman to P. H. Sheridan, W. B. Hazen, and B. H. Grierson, December 23, 1868, Roll 14, General Correspondence, Sherman Papers, and "Sheridan and the Army," *Army and Navy Register* (hereinafter *ANR*) 8 (December 3, 1887), 773.

5. Report of Headquarters Post Fort Philip Kearney, January 3, 1867, Roll 560, Fetterman Papers, AGO File 102 M 1867, M-619, National Archives; AR, Report of the Department of the Dakota, October 5, 1868, Roll 16, 35; John G. Bourke, *On the Border with Crook* (Lincoln, Nebr., 1971), 107. See also "Apaches Still on the War Path," *ANR* 8 (July 16, 1887), 451. Joseph C. Porter, *Paper Medicine Man: John Gregory Bourke and His American West* (London and Norman, Okla., 1986), is a well-crafted biography that does justice to Bourke's career as both soldier and ethnologist.

6. Report of the Department of Arizona, July 23, 1883, Roll 174, Chiricahua Apache Papers, File 1066 AGO 1883, M-689, National Archives; Stephen C. Mills to My Dear Mother, May 9, 1882, Stephen C. Mills Papers; AR, Report of the Military Division of the Pacific, October 20, 1871, Roll 18, 67; George Crook to John M. Schofield, May 9, 1873, Box 39, Special Correspondence, John M. Schofield Papers, Library of Congress; "Apaches Still on the War Path," *ANR* 8 (July 16, 1887), 451; Keith Clark and Donna Clark, eds., "William McKay's Journal, 1866–67: Indian Scouts, Part I," *Oregon Historical Quarterly* 79 (Summer 1978), 152.

7. AR, Report of the Division of the Missouri, October 9, 1885, Roll 50, 128; AR, Report of the Department of the Columbia, September 18, 1880, Roll 35, 190; Report of Lieutenant General P. H. Sheridan, Military Division of the Missouri, October 1, 1874, Military Division of the Missouri Papers, U.S. Army Military History Institute. This quotation also appears on AR, Report of the Military Division of the Missouri, October 1, 1874, Roll 22, 23. See also Sheridan's comments the preceding year, Report of the Military Division of Missouri, October 27, 1873, Roll 20, 40.

8. Utley, *Frontier Regulars*, 4–5, 148–149, 221; Report of the Department of the Missouri, April 23, 1867, Roll 563, Reports of Campaigns of Generals Hancock and Custer against the Sioux and Cheyenne, AGO File 590 M 1867, M-619, National Archives. See also William T. Sherman to W. S. Hancock, March 14, 1867, ibid.

9. U. S. Grant to William T. Sherman, March 12, 1866, Roll 10, General Correspondence, Sherman Papers.

10. William T. Sherman to P. H. Sheridan, W. B. Hazen, and B. H. Grierson, December 23, 1868, Roll 14, General Correspondence, Sherman Papers; AR, Report of the Department of the Missouri, October 31, 1870, Roll 17, 10; AR, Report of the Department of Dakota, October 3, 1872, Roll 19, 41.

11. E. F. Townsend to Luther P. Bradley, June 4, 1879, Bradley Family Letters, Luther P. Bradley Papers, U.S. Army Military History Institute; AR, Report of the District of the Yellowstone, September [n.d.] 1879, Roll 33, 73; AR, Report of the Department of the Missouri, October 3, 1879, 85; AR, Report of the District of the Pecos, September 20, 1880, quoted in the Report of the Department of Texas, October 1, 1880, Roll 35, 63; quoted in Bourke, *On the Border*, 467.

12. AR, Report of the District of the Yellowstone, September [n.d.] 1879, Roll 33, 73; George A. Custer to P. H. Sheridan, November 28, 1868, Military Division of the Missouri Papers; Telegram, George Crook to the Adjutant General, U.S. Army, June 12, 1883, Roll 173, Chiricahua Apache Papers. This telegram is unsigned, but it is certain Crook was the author. See Assistant Adjutant General, Department of Arizona, to the Adjutant General, U.S. Army, June 12, 1883, Roll 173, Chiricahua Apache Papers.

13. John G. Bourke's Diary, 67:24, quoted in Dan L. Thrapp, *General Crook and the Sierra Madre Adventure* (Norman, Okla., 1972), 139.

14. Entry for August 20, 1867, Private Journal of Luther P. Bradley, 1867–1880, Luther P. Bradley Papers; AR, Report of the Department of Arizona, August 31, 1874, Roll 22, 61.

15. William T. Sherman to Nelson A. Miles, August 23, 1876, Box 4, Nelson

A. Miles Family Papers, Library of Congress; Nelson A. Miles to William T. Sherman, January 20, 1877, Roll 23, General Correspondence, Sherman Papers.

16. De B. Randolph Keim, *Sheridan's Troopers on the Borders: A Winter Campaign on the Plains* (London and Lincoln, Nebr., 1985), 174; George A. Custer to P. H. Sheridan, November 28, 1868, Military Division of the Missouri Papers; Guy V. Henry, "Adventures of American Army and Navy Officers: A Winter March to the Black Hills," undated article from *Harper's Weekly* in the Guy V. Henry Papers, U.S. Army Military History Institute.

17. John G. Bourke's Diary, quoted in J. W. Vaughn, *The Reynolds Campaign on Powder River* (Norman, Okla., 1961), 53; Bourke, *On the Border*, 263. See also Autobiography of Major General George Crook, U.S.A., 103, Crook-Kennon Papers, U.S. Army Military History Institute, and Nelson A. Miles to William T. Sherman, December 25, 1876, Roll 23, General Correspondence, Sherman Papers.

18. Col. Joseph J. Reynolds, Report on the Big Horn Expedition, April 15, 1876, in Vaughn, *Reynolds Campaign*, 213.

19. Bourke, *On the Border*, 275–276; J. J. Reynolds to William T. Sherman, April 11, 1876, Roll 21, General Correspondence, Sherman Papers.

20. Report of Bvt. Maj. Gen. Nelson A. Miles, Indian Territory Expedition, March 4, 1875, Military Division of the Missouri Papers; Report of Major A. P. Morrow, Battalion Ninth Cavalry, January 23, 1875, ibid.; Martin F. Schmitt, ed., *General George Crook: His Autobiography* (London and Norman, Okla., 1986), 206, 209; Stephen C. Mills to the Regimental Adjutant, Twelfth U.S. Infantry, July 12, 1907, Stephen C. Mills Papers. See also Utley, *Frontier Regulars*, 228.

The problem of logistics was answered in part by railroads. See Utley, *Frontier Regulars*, 93–94; Robert G. Athearn, "The Firewagon Road," in Paul L. Hedren, ed., *The Great Sioux War, 1876–77* (Helena, Mont., 1991), 65–84; and AR, Report of the Commanding General, October 27, 1883, Roll 44, 45–46. It was also addressed by the use of mule packtrains, a particular interest of George Crook, who by 1874 had formed the opinion that "American horses" needed at least a year to become "acclimated" to service in the Arizona Territory. George Crook to John M. Schofield, June 12, 1874, Box 39, Special Correspondence, Schofield Papers. On Crook's use of mules, see Utley, *Frontier Regulars*, 48–49; Emmett M. Essin III, "Mules, Packs, and Packtrains," *Southwestern Historical Quarterly* 74 (July 1970), 52–80; "General Crook Speaks," *ANJ* 14 (October 31, 1876), 166; AR, Report of the Department of Arizona, April 10, 1886, Roll 53, 149; Bourke, *On the Border*, 150–154, 187. See also Philip H. Sheridan to the Secretary of War, July 6, 1885, Box 35, General Correspondence, Philip H. Sheridan Papers, Library of

Congress, and AR, Report of the Division of the Missouri, October 9, 1885, Roll 50, 125 and 128.

21. Stephen C. Mills to the Regimental Adjutant, Twelfth U.S. Infantry, July 12, 1907, Stephen C. Mills Papers; W. H. Timmons, ed., *Four Centuries at the Pass* (El Paso, 1980), 82; entry for August 30, 1873, Private Journal of Luther P. Bradley, 1867–1880, Luther P. Bradley Papers. See also the entry for the same date in the Private Journal of Luther P. Bradley, 1873–1874, ibid.

22. Utley, *Frontier Regulars*, 388.

23. William T. Sherman to Henry Ward Beecher, March 6, 1879, Roll 46, Letterbooks, Sherman Papers; Nelson A. Miles, *Serving the Republic: Memoirs of the Civil and Military Life of Nelson A. Miles* (Freeport, N.Y., 1971), 117–118 and 163; Nelson A. Miles to William T. Sherman, November 1, 1876, Roll 23, General Correspondence, Sherman Papers. On Indian arms, see "Sheridan and the Army," *ANR* 8 (December 3, 1887), 773.

24. Bourke, *On the Border*, 247; Anson Mills and C. H. Claudy, ed., *My Story* (Washington, D.C., 1918), 406; AR, Report of Major Marcus A. Reno, July 5, 1876, quoted in the Report of the General of the Army, November 10, 1876, Roll 26, 33.

25. Report of Brevet Major General G. A. Custer, Troops Operating South of the Arkansas, March 21, 1869, Military Division of the Missouri Papers; AR, Report of the Department of the Missouri, October 31, 1870, Roll 17, 10; "A Fight with the Comanches and Kiowas," *ANJ* 12 (October 31, 1874), 187. On the horsemanship of the Plains Indians, see also Sherry L. Smith, *The View from Officers' Row: Army Perceptions of Western Indians* (Tucson, 1990), 149–150.

26. Edward S. Farrow, *Mountain Scouting: A Handbook for Officers and Soldiers on the Frontiers* (New York, 1881), 240; Mills, *My Story*, 406; Nelson A. Miles to William T. Sherman, November 1, 1879, Roll 23, General Correspondence, Sherman Papers. See also AR, Report of the Department of Arizona, September 27, 1883, Roll 44, 166.

27. AR, Report of the General of the Army, November 1, 1879, Roll 33, 7; AR, Report of the Military Division of the Pacific, September 18, 1867, 71; William T. Sherman quoted in Charles D. Rhodes, "Chief Joseph and the Nez Percés Campaign of 1877: An Epic of the American Indian," in John M. Carroll, ed., *The Papers of the Order of the Indian Wars* (Fort Collins, Colo., 1975), 231; John Gibbon to Nelson A. Miles, October 21, 1877, Box 2, Nelson A. Miles Family Papers.

28. Utley, *Frontier Regulars*, 198; William T. Sherman to Herbert A. Preston, April 17, 1873, Roll 44, Letterbooks, Sherman Papers; Philip H. Sheridan to William T. Sherman, June 6, 1873, Roll 19, General Correspondence, ibid.

29. Bourke, *On the Border*, 467–468; Report of the Department of Arizona, July 23, 1883, Roll 174, Chiricahua Apache Papers; Report on Operations against the Chiricahuas, January 11, 1886, Roll 181, ibid.

30. William T. Sherman to Philip H. Sheridan, February 17, 1877, Roll 45, Letterbooks, Sherman Papers; Charles King, *Proceedings of the Annual Meeting and Dinner of the Order of the Indian Wars of the United States*, February 26, 1921, File S-4, Order of the Indian Wars Papers, U.S. Army Military History Institute. See also Utley, *Frontier Regulars*, 262.

31. Utley, *Frontier Regulars*, 105–106; Report of Headquarters Post Fort Philip Kearny, January 3, 1867, Roll 560, Fetterman Papers. See also Grace Raymond Hebard and E. A. Brininstool, *The Bozeman Trail*, 2 vols. (Cleveland, 1922), 1:297–343.

32. Utley, *Frontier Regulars*, 201–202; Vaughn, *Reynolds Campaign*, 187.

33. Utley, *Frontier Regulars*, 301.

34. William T. Sherman to Philip H. Sheridan, February 17, 1877, Roll 45, Letterbooks, Sherman Papers; George Crook to Philip H. Sheridan, July 16, 1876, Roll 271, Rosebud Papers, File 3570 AGO 1876, M-666, National Archives; George Crook to Philip H. Sheridan, July 23, 1876, Roll 271, ibid. For assessments of the Rosebud as an Indian victory, see Neil C. Mangum, *Battle of the Rosebud: Prelude to the Little Big Horn* (El Segundo, Calif., 1987), 91; Utley, *Frontier Regulars*, 256, and Schmitt, *Crook Autobiography*, 196.

35. Report of Lt. Col. W. B. Royall, Third Cavalry, June 20, 1876, Roll 271, Rosebud Papers; Farrow, *Mountain Scouting*, 240; Jerome A. Greene, *Slim Buttes, 1876: An Episode of the Great Sioux War* (Norman, Okla., 1982), 74; John F. Finerty, *War-Path and Bivouac: The Big Horn and Yellowstone Expedition* (London and Lincoln, Nebr., 1966), 284. On the Indians' use of conventional tactics at the Rosebud, see also Mangum, *Battle of the Rosebud*, 93. For other examples of Indians using entrenchments, see AR, Report of Lt. Col. Thomas H. Neill, Commanding Sixth Cavalry, April 7, 1875, Roll 24, 87–88, and Report of Lt. W. P. Clark, Second Cavalry, quoted in the Report of the Department of Dakota, October 1, 1879, Roll 33, 57.

36. Rhodes, "Chief Joseph," in Carroll, ed., *Order of the Indian Wars*, 231.

37. Farrow, *Mountain Scouting*, 239; Randolph B. Marcy, *Thirty Years of Army Life on the Border* (Philadelphia and New York, 1963), 48.

38. Marcy, *Thirty Years of Army Life,* 47–48. See also AR, Report of the Department of Arizona, September 27, 1883, Roll 44, 167, and Rickey, *Forty Miles a Day,* 282–283.

39. John G. Bourke, *An Apache Campaign in the Sierra Madre* (New York, 1958), 27–28; AR, Report of the Department of Arizona, August 31, 1882, Roll 41, 151; George Bird Grinnell, *The Fighting Cheyennes* (Norman, Okla., 1971), 5.

40. Edward S. Godfrey, "Custer's Last Battle," 1908, Godfrey Family Papers, U.S. Army Military History Institute; W. S. Edgerly, in "Foreword" to W. A. Graham, *The Story of the Little Big Horn: Custer's Last Fight* (London and Lincoln, Nebr., 1988), xxxvi; "General Hugh L. Scott Discusses the Sioux War," in W. A. Graham, *The Custer Myth: A Source Book of Custeriana* (New York, 1953), 114; Edgar I. Stewart, *Custer's Luck* (Norman, Okla., 1985), 317, 319; Robert M. Utley, *Cavalier in Buckskin: George Armstrong Custer and the Western Military Frontier* (London and Norman, Okla., 1988), 180, 182.

3. The Same Principle as at Atlanta

1. Robert M. Utley, *Frontier Regulars: The United States Army and the Indian, 1866–1891* (New York and London, 1973), 36; Thomas W. Dunlay, *Wolves for the Blue Soldiers: Indian Scouts and Auxiliaries with the United States Army, 1860–90* (London and Lincoln, Nebr., 1982), 73; Paul Andrew Hutton, *Phil Sheridan and His Army* (London and Lincoln, Nebr., 1985), 145. Robert Wooster accepted Hutton's thinking on this point: *The Military and United States Indian Policy, 1865–1903* (London and New Haven, Conn., 1988), 171.

2. Dunlay, *Wolves for the Blue Soldiers,* 74, 76, 236 n. 21.

3. Larry D. Roberts, U.S. Army Combat Studies Institute, "Strategy, Doctrine, and the Frontier Army" (MS in possession of P. D. Jamieson).

4. Hutton, *Sheridan and His Army,* 54. This point of view is consistent with that taken in Utley, *Frontier Regulars,* 150. On the winter-war strategy, see also Hutton, ibid., 54–55; Russell F. Weigley, *The American Way of War: A History of United States Military Strategy and Policy* (New York and London, 1976), 159; and Roy Morris, Jr., *Sheridan: The Life and Wars of General Phil Sheridan* (New York, 1992), 307–308. For two examples of contemporary thinking, see Philip H. Sheridan to William T. Sherman, February 10, 1877, Roll 23, General Correspondence, William T. Sherman Papers, Library of Congress, and "General Crook Speaks," *ANJ* 14 (October 31, 1876), 166.

5. Gibbon quoted in Jerome A. Greene, *Yellowstone Command: Colonel Nelson A. Miles and the Great Sioux War, 1876–1877* (London and Lincoln, Nebr., 1991), 70; John G. Bourke, *On the Border with Crook* (Lincoln, Nebr., 1971), 251; AR, Report of the District of Arizona, April 29, 1866, Roll 15, 97.

6. Utley, *Frontier Regulars*, 149–150, map, 146; Hutton, *Sheridan and His Army*, 53–54, map, 58; Morris, *Sheridan*, 310. On the converging-columns strategy, see also Wooster, *The Military and Indian Policy*, 32, 40.

7. Utley, *Frontier Regulars*, 220–221, map, 222; Hutton, *Sheridan and His Army*, 248–249, map, 250.

8. Utley, *Frontier Regulars*, 251–252; Hutton, *Sheridan and His Army*, 303; Robert M. Utley, *Cavalier in Buckskin: George Armstrong Custer and the Western Military Frontier* (London and Norman, Okla., 1988), 169–170. As John S. Gray and other historians have pointed out, there was no expectation that the Custer and Terry-Gibbon columns—let alone the Crook column far to the south—would converge at a predetermined location on a predetermined date. John S. Gray, *The Centennial Campaign: The Sioux War of 1876* (London and Norman, Okla., 1988), 142–147.

9. William T. Sherman to John Sherman, May 18, 1871, Roll 16, General Correspondence, Sherman Papers; Ranald S. Mackenzie to C. C. Augur, June 28, 1873, Letterbook, 1873–1874, Ranald S. Mackenzie Papers, U.S. Army Military History Institute; P. T. Swaine to William T. Sherman, August 24, 1881, Roll 29, General Correspondence, Sherman Papers; Report of Operations against the Chiricahuas, January 11, 1886, Roll 181, Chiricahua Apache Papers, File 1066 AGO 1883, M-689, National Archives. See also Gray, *The Centennial Campaign*, 321.

10. George Crook to John M. Schofield, September 1, 1871, Box 39, Special Correspondence, John M. Schofield Papers, Library of Congress; "General Crook Speaks," 166; AR, Report of the Department of Arizona, September 27, 1883, Roll 44, 166; John K. Mahon, *History of the Second Seminole War, 1835–1842* (Gainesville, 1985), 324 and 325.

11. Mahon, *Second Seminole War*, 172–173, 180, 182–184, 195–196; Douglas Edward Leach, *Arms for Empire: A Military History of the British Colonies in North America, 1607–1763* (New York and London, 1973), 64–66 and passim; Douglas Edward Leach, *Flintlock and Tomahawk: New England in King Philip's War* (New York, 1966), 145–154 and passim; Dunlay, *Wolves for the Blue Soldiers*, 77. Dunlay's book is a thoughtful treatment of the post–Civil War Indian scouts and auxiliaries, and there is also a sound assessment of white attitudes about them in Sherry L. Smith, *The View from Officers' Row: Army Perceptions of Western Indians* (Tuc-

son, 1990), 163–181. On the service of scouts against the Apaches in particular, see Joyce Evelyn Mason, "The Use of Indian Scouts in the Apache Wars, 1870–1886" (Ph.D. diss., Indiana University, 1970).

12. Neil C. Mangum, *Battle of the Rosebud: Prelude to the Little Big Horn* (El Segundo, Calif., 1987), 54–55; Report of Lt. Col. W. B. Royall, June 20, 1876, Roll 271, Rosebud Papers, File 3570 AGO 1876, M-666, National Archives.

13. John S. Gray, *Custer's Last Campaign: Mitch Boyer and the Little Big Horn Reconstructed* (London and Lincoln, Nebr., 1991), 200. The spelling of "Boyer" follows Gray, his most thorough biographer. On the Crow scouts, see also Edgar I. Stewart, *Custer's Luck* (Norman, Okla., 1985), 245, 246, and Utley, *Cavalier in Buckskin*, 176.

14. Dunlay, *Wolves for the Blue Soldiers*, 82–84. On North and the Pawnee scouts, see also George Bird Grinnell, *Two Great Scouts and Their Pawnee Battalion* (Cleveland, 1928).

15. AR, Report of Maj. Marcus A. Reno, July 5, 1876, quoted in the Report of the General of the Army, November 10, 1876, Roll 26, 34; AR, Report of the Department of the Dakota, October 20, 1869, Roll 17, 66; George Crook to John M. Schofield, September 1, 1871, Box 39, Special Correspondence, John M. Schofield Papers; P. T. Swaine to William T. Sherman, August 24, 1881, Roll 29, General Correspondence, Sherman Papers; AR, Report of Capt. Guido Ilges, April 30, 1867, Roll 15, 55; Dunlay, *Wolves for the Blue Soldiers*, 170–173.

16. Dunlay, *Wolves for the Blue Soldiers*, 165.

17. AR, Report of the Lieutenant General of the Army, October 10, 1886, Roll 53, 72; John C. Kelton to George Crook, March 8, 1887, Roll 188, Chiricahua Apache Papers.

18. R. C. Drum to the Commanding General, Department of the Platte, April 20, 1887, Roll 188, Chiricahua Apache Papers; George Crook to Philip H. Sheridan, April 7, 1886, Roll 188, ibid.; Bourke, *On the Border*, 203, 468; AR, Report of the Military Division of the Pacific, September 22, 1868, Roll 16, 48.

19. AR, Report of the Department of the Platte, September 30, 1867, Roll 15, 60; C. C. Augur to William T. Sherman, March 10, 1869, Roll 14, General Correspondence, Sherman Papers; Miles, *Serving the Republic*, 179; Report of Bvt. Maj. Gen. Nelson A. Miles, Indian Territory Expedition, March 4, 1875, Military Division of the Missouri Papers, U.S. Army Military History Institute; R. H. Pratt to William T. Sherman, October 31, 1876, Roll 23, General Correspondence, Sherman Papers; Dunlay, *Wolves for the Blue Soldiers*, 51. See also ibid., 50–54; Wooster, *The Military and Indian Policy*, 127–128; AR, Report of the Military Division of the Pacific, September 22, 1868, Roll 16, 48; AR, Report of the Depart-

ment of Arizona, August 15, 1877, Roll 28, 148; AR, Report of the Secretary of War, November 19, 1878, Roll 30, iv; AR, Report of the Department of the Columbia, September 18, 1880, Roll 35, 189–190.

20. E. S. Godfrey, "Some Reminiscences," _Cavalry Journal_ 36 (July 1927), 421, File R-6, Order of the Indian Wars Papers, U.S. Army Military History Institute; "General Godfrey's Narrative," in W. A. Graham, _The Custer Myth: A Source Book of Custeriana_ (New York, 1953), 125 and 127; De B. Randolph Keim, _Sheridan's Troopers on the Borders: A Winter Campaign on the Plains_ (London and Lincoln, Nebr., 1985), 94.

21. "Commanches and Kiowas," _ANJ_ 12 (October 31, 1874), 186; Francis B. Heitman, _Historical Register and Dictionary of the United States Army,_ 2 vols. (Washington, D.C., 1903), 1:630, 648.

22. James T. Kerr, "The Modoc War of 1872–73," in John M. Carroll, ed., _Order of the Indian Wars_ (Fort Collins, Colo., 1975), 145.

23. Report of Maj. Alexander Chambers, June 20, 1876, Roll 271, Rosebud Papers; John F. Finerty, _War-Path and Bivouac: The Big Horn and Yellowstone Expedition_ (London and Lincoln, Nebr., 1966), 127; Neil C. Mangum, _Battle of the Rosebud: Prelude to the Little Big Horn_ (El Segundo, Calif., 1987), 57.

24. Report of Lt. Col. W. B. Royall, June 20, 1876, Roll 271, Rosebud Papers; Mangum, _Battle of the Rosebud,_ 78.

25. Frank U. Robinson, "The Battle of Snake Mountain," in Carroll, _Order of the Indian Wars,_ 188; Martin F. Schmitt, ed., _General George Crook: His Autobiography_ (London and Norman, Okla., 1986), 45. For an example of skirmishers protecting a military wagon train, see Greene, _Yellowstone Command,_ 88.

26. Grady McWhiney and Perry D. Jamieson, _Attack and Die: Civil War Military Tactics and the Southern Heritage_ (University, Ala., 1982), 99–101; AR, Report of the Department of Arizona, September 27, 1883, Roll 44, 166; John Gibbon, "Arms to Fight Indians," _United Service_ 1 (April 1879), 240; Charles King, _Campaigning With Crook_ (Norman, Okla., 1964), 120, 122 (quotations).

27. Jerome A. Greene, _Slim Buttes, 1876: An Episode of the Great Sioux War_ (Norman, Okla., 1982), 57–58; Utley, _Frontier Regulars,_ 155–157, 208, 256 and 258.

28. Greene, _Slim Buttes,_ 58; G. O. Shields, _The Battle of the Big Hole_ (Chicago and New York, 1899), 47–48; Utley, _Frontier Regulars,_ 312; Joseph C. Porter, _Paper Medicine Man: John Gregory Bourke and His American West_ (London and Norman, Okla., 1986), 55. See also the maps on 31 and 55 of Fred H. Werner, _The Dull Knife Battle_ (Greeley, Colo., 1981), which correct the compass points on the map in Grinnell, _The Fighting Cheyennes,_ 366.

29. Edward S. Godfrey, "General Godfrey's Narrative," in Graham, *Custer Myth*, 147.

30. William T. Sherman to Ulysses S. Grant, May 27, 1867, Roll 563, Union Pacific Railroad File, AGO File 597 M 1867, M-619, National Archives; H. Douglas to Chauncey McKeever, June 18, 1867, Roll 560, Fetterman Papers, AGO File 102 M 1867, ibid.; Thomas L. Rosser to the St. Paul and Minneapolis *Pioneer-Press Tribune*, July 8, 1876, in Graham, *Custer Myth*, 225; E. J. McClernand, "With the Indian and the Buffalo in Montana," *Cavalry Journal* 36 (April 1927), 192; Nelson A. Miles to William T. Sherman, October 23, 1876, Roll 23, General Correspondence, Sherman Papers. See also "Sheridan and the Army," *ANR* 8 (December 3, 1887), 773.

31. Greene, *Yellowstone Command*, 112; Nelson A. Miles to William T. Sherman, June 15, 1876, Roll 23, General Correspondence, Sherman Papers; William T. Sherman to John M. Schofield, August 6, 1878, Roll 45, Letterbooks, ibid.; Philip H. Sheridan to William T. Sherman, April 5, 1878, ibid. See also James S. Hutchins, "Mounted Riflemen: The Real Role of Cavalry in the Indian Wars," *El Palacio* 69 (Summer 1962), 85–91.

32. Don Rickey, Jr., *Forty Miles a Day on Beans and Hay: The Enlisted Soldier Fighting the Indian Wars* (Norman, Okla., 1963), 219; Utley, *Frontier Regulars*, 73; Col. Henry J. Hunt to Subcommittee on Reorganization of the Army, House Committee on Military Affairs, February 11, 1878, House Miscellaneous Document No. 56, 45th Cong., 2d sess., 95.

33. Col. Henry J. Hunt to Subcommittee on Reorganization of the Army, House Committee on Military Affairs, February 11, 1878, 95; Nelson A. Miles to William T. Sherman, July 8, 1876, Roll 23, General Correspondence, Sherman Papers.

34. AR, Report of the Department of the Columbia, August 27, 1877, Roll 28, 122.

35. Utley, *Frontier Regulars*, 303; Charles D. Rhodes, "Chief Joseph and the Nez Percés Campaign of 1877: An Epic of the American Indian," in Carroll, *Order of the Indian Wars*, 219; AR, Report of the Department of the Columbia, August 27, 1877, Roll 28, 124.

36. Bvt. Maj. J. M. Williams, "Indian Scouting in Arizona in 1867," undated typescript, J. M. Williams Papers, Military Order of the Loyal Legion of the United States Collection, U.S. Army Military History Institute; Nelson A. Miles, *Personal Recollections and Observations of General Nelson A. Miles* (Chicago and New York, 1896), 228; Nelson A. Miles to William T. Sherman, January 20, 1877,

Roll 23, General Correspondence, Sherman Papers. Historian Don Rickey, Jr., concluded that the Indians had "greatly feared" artillery on other battlefields, but at the Wolf Mountains the guns did not alone "tip the scales of battle." Don Rickey, Jr., "The Battle of Wolf Mountain," in Paul L. Hedren, ed., *The Great Sioux War, 1876–77* (Helena, Mont., 1991), 170. On Miles and the artillery, see also Greene, *Yellowstone Command,* 76. On artillery during the Indian wars, see also S. E. Whitman, *The Troopers: An Informal History of the Plains Cavalry, 1865–1890* (New York, 1962), 185–186, and Wilbur Sturtevant Nye, *Plains Indian Raiders: The Final Phases of Warfare from the Arkansas to the Red River* (Norman, Okla., 1974), 161.

37. AR, Report of Lt. Col. Thomas H. Neill, April 7, 1875, Roll 24, 87; Nelson A. Miles, *Serving the Republic: Memoirs of the Civil and Military Life of Nelson A. Miles* (Freeport, N.Y., 1971), 151–152; Robert M. Utley, *The Last Days of the Sioux Nation* (London and New Haven, Conn., 1963), 216–222. See also Utley, *Frontier Regulars,* 406.

38. Utley, *Frontier Regulars,* 46–47, 95, maps, 96, 165.

39. Wooster, *The Military and Indian Policy,* 163–164.

40. Ibid., 124–125, 148, 213, 260–261, 336–337. On the Wagon Box and Hayfield fights, see Grace Raymond Hebard and E. A. Brininstool, *The Bozeman Trail,* 2 vols. (Cleveland, 1922), 2:39–87 and 2:159–174.

41. Utley, *Frontier Regulars,* 51–52, 144–145; Russell F. Weigley, *The American Way of War: A History of United States Military Strategy and Policy* (New York and London, 1976), 153ff.; Wooster, *The Military and Indian Policy,* 4, 142–143.

42. Ibid., 28–29 and 40 (quotation, 29).

43. Miles, *Serving the Republic,* 122; Report of Bvt. Maj. Gen. Nelson A. Miles, Indian Territory Expedition, March 4, 1875, Military Division of the Missouri Papers. See also Miles's account of his August 30, 1874, engagement against a Cheyenne, Commanche, and Kiowa war party, in this same report.

44. Miles, *Serving the Republic,* 163.

45. AR, Report of the Military Division of the Pacific, October 18, 1866, Roll 15, 32; AR, Report of the Department of California, September 14, 1867, Roll 15, 128; Grant quoted in Wooster, *The Military and Indian Policy,* 127; Ranald S. Mackenzie to C. C. Augur, June 2, 1873, Letterbook, 1873–1874, Ranald S. Mackenzie Papers; Ranald S. Mackenzie to C. C. Augur, June 28, 1873, ibid. See also Ranald S. Mackenzie to C. C. Augur, July 17, 1873, ibid.

46. Philip H. Sheridan to George Crook, June 9, 1885, Box 35, General Correspondence, Philip H. Sheridan Papers, Library of Congress. See also AR, Report of the Lieutenant General of the Army, October 24, 1885, Roll 50, 61.

47. Report of Headquarters Post Fort Philip Kearny, January 3, 1867, Roll 560, Fetterman Papers; William T. Sherman to P. H. Sheridan, W. B. Hazen, and B. H. Grierson, December 23, 1868, Roll 14, General Correspondence, Sherman Papers; "General Crook Speaks," 166; Philip H. Sheridan to William T. Sherman, December 12, 1879, Roll 26, General Correspondence, Sherman Papers.

48. Hutton, *Sheridan and His Army*, 50, 54–55, 63, 190, 248–249, 301; P. H. Sheridan to William T. Sherman, January 20, 1869, and June 6, 1873, Rolls 14 and 19, General Correspondence, Sherman Papers; Philip H. Sheridan to the Honorable Secretary of War, December 28, 1883, Box 34, General Correspondence, Sheridan Papers.

49. John Pope to William T. Sherman, September 16, 1874, Roll 20, General Correspondence, Sherman Papers.

50. George A. Custer to P. H. Sheridan, November 28, 1868, Military Division of the Missouri Papers; Custer quoted in Utley, *Cavalier in Buckskin*, 70.

51. John F. Finerty, "Life on the Frontier," *Citizen*, April 22, 1893, newspaper clipping in the Guy V. Henry Papers, U.S. Army Military History Institute; U.S. War Department, *The War of the Rebellion: A Compilation of the Official Records of the Union and Confederate Armies,* 128 vols. (Washington, D.C., 1880–1901), ser. 1, 30, pt. 1, 175, 594–595; entries for September 19, 1867, and September 19, 1868, Private Journal of Luther P. Bradley, 1867–1880, Luther P. Bradley Papers, U.S. Army Military History Institute; Miles quoted in Utley, *Frontier Regulars*, 228. See also Greene, *Yellowstone Command*, 22.

52. Utley, *Frontier Regulars*, 144; Weigley, *American Way of War*, 159. Morris, *Sheridan*, 307, agreed with Utley and Weigley that the total war strategy of the Indian wars derived from the Civil War. Greene also followed Utley and Weigley on this subject while acknowledging Wooster's "alternative explanation": Greene, *Yellowstone Command*, 12 and 239.

53. Wooster, *The Military and Indian Policy*, 136–141.

54. Ibid., 142–143.

55. Ibid., 111–143, 210, 213, 216.

56. Ibid., 111, 135–136.

57. On the Civil War careers of Indian-fighting commanders, see Wooster, *The Military and Indian Policy*, 55–56 and 135. Northern veterans did not acknowledge that their successful offensives in 1864 and 1865 had been possible because Southern commanders had bled their own forces so badly by reckless assaults during the first three years of the war. McWhiney and Jamieson, *Attack and Die*, 7 and 9.

58. Ulysses S. Grant to William T. Sherman, September 25, 1868, Roll 13, General Correspondence, Sherman Papers; Sheridan quoted in Hutton, *Sheridan and His Army*, 185; Nelson A. Miles to William T. Sherman, January 8, 1878, Roll 24, General Correspondence, Sherman Papers; William T. Sherman to T. C. Durant, May 28, 1867, Union Pacific Railroad File.

4. *Individual Skill and the Hazard of Battle*

1. AR, Report of the Department of the Missouri, October 2, 1883, Roll 44, 1:136; AR, Report of Capt. H. C. Cushing, July 25, 1883, quoted in the Report of the Adjutant General, October 30, 1883, ibid., 348; AR, Report of the Division of the Missouri, September 10, 1886, Roll 53, 120; Col. John C. Kelton, "Observations," August 8, 1881, AGO File 1558 AGO 1882, Recommendations of Lt. Col. James W. Forsyth, Roll 94, M-689, National Archives. See also AR, Report of the Lieutenant-General of the Army, November 1, 1884, Roll 47, 48.

2. AR, Report of the Department of the Missouri, September 22, 1881, Roll 38, 125; John Gibbon, "Arms to Fight Indians," *United Service* 1 (April 1879), 240; E. S. Godfrey, "Cavalry Fire Discipline," *JMSIUS* 19 (July–November 1896), 252–259; T. J., "From Fort Keough, Montana," *ANR* 3 (July 15, 1882), 3. See also H. G. Litchfield, "Military Rifles and Rifle Firing: Marksmanship as an Element of National Strength," *JMSIUS* 1 (1880), 302–303. For the perspectives of three historians on this subject, see Don Rickey, Jr., *Forty Miles a Day on Beans and Hay: The Enlisted Soldier Fighting the Indian Wars* (Norman, Okla., 1963), 280; for Rickey's discussion of training and target practice in particular, see 102–105; Russell Gilmore, "'The New Courage': Rifles and Soldier Individualism, 1876–1918," *Military Affairs* 40 (October 1976), 98; Robert M. Utley, *Frontier Regulars: The United States Army and the Indian, 1866–1891* (New York and London, 1973), 24–25. Gilmore's well-crafted article makes clear the connection between the army's attention to marksmanship and its emphasis on individualism.

3. Gilmore, "'The New Courage,'" 97; Donald N. Bigelow, *William Conant Church and the Army Navy Journal* (New York, 1952), 184–185; George W. Wingate, *Manual for Rifle Practice* (New York, 1875). T. S. S. Laidley, *A Course of Instruction in Rifle Firing* (Philadelphia, 1879), which copied Wingate's efforts, eventually became the standard work on marksmanship. See Gilmore, "'The New Courage,'" 97.

4. AR, Report of the Inspector-General, October 1, 1880, Roll 35, 172; Gilmore, "'The New Courage,'" 97.

5. Matthew F. Steele, "Military Reading: Its Use and Abuse," *JUSCA* 8 (1895), 93; Eben Swift, "Sabers or Revolvers?," *JUSCA* 1 (1888), 49; Stephen C. Mills to the Regimental Adjutant, Twelfth U.S. Infantry, July 12, 1907, Stephen C. Mills Papers, U.S. Army Military History Institute. See also "National Rifle Association," *ANR* 3 (April 8, 1882), 12; J. M. T. Partello, "Rifle Practice in the Army," *ANR* 3 (April 22, 1882), 4; "The Rifle Match Report," *ANR* 4 (September 8, 1883), 11.

6. AR, Report of the Lieutenant-General of the Army, October 24, 1885, Roll 50, 64. See also AR, Report of Secretary of War, November 30, 1885, Roll 50, 7.

7. Litchfield, "Military Rifles," 307; AR, Report of the Division of the Atlantic, October 7, 1885, Roll 50, 121; AR, Report of the Inspector General of the Army, October 10, 1888, Roll 58, 109. See also AR, Report of the Lieutenant General of the Army, October 10, 1886, Roll 53, 75–76, and Stephen E. Ambrose, *Upton and the Army* (Baton Rouge, 1964), 141.

8. AR, Col. Robert P. Hughes quoted in the Report of the Inspector General of the Army, October 1, 1890, Roll 64, 107; AR, Report of the Inspector General of the Army, October 10, 1888, Roll 58, 108; Stanhope E. Blunt, *Instructions in Rifle and Carbine Firing for the United States Army* (New York, 1885); Major Robert H. Hall quoted in AR, Report of the Inspector General of the Army, October 6, 1886, Roll 53, 113; Lt. Col. Henry C. Hasbrouck, quoted in AR, Report of the Inspector General of the Army, October 18, 1897, Roll 93, 134.

9. AR, Report of the Inspector General of the Army, October 1, 1890, Roll 64, 108; ibid., September 27, 1895, Roll 85, 114; "Target Practice," *ANJ* 21 (October 27, 1883), 239; Moses Harris to James W. Forsyth, February 10, 1882, File 1558 AGO 1882, Recommendations of Lt. Col. James W. Forsyth; John W. Ruckman, "Artillery Difficulties in the Next War," *JUSA* 2 (1893), 436, 437. On infantry target practice, see also G.W.W., "Military Rifle Shooting," *ANJ* 21 (November 3, 1883), 276, and on artillery target practice, see also note 25.

10. C. D. Parkhurst, "The Practical Education of the Soldier," *JMSIUS* 11 (1890), 952; August V. Kautz, "What the Army Should Be," *ANR* 13 (February 6, 1892), 92; "Forward," "Thoughts Pertinent to the Introduction of Our New Infantry Drill Regulations," *ANR* 13 (March 5, 1892), 157; William T. Sherman, address to the Class of 1879, Michigan Military Academy, June 19, 1879, Roll 45, Letterbooks, Sherman Papers.

11. AR, Report of the U.S. Infantry and Cavalry School, July 1, 1891, Roll 68, 265–266; AR, Report of the Commandant of Cadets, August 28, 1891, Roll 68, 293. See also, AR, Report of the Superintendent of the U.S. Military Academy, September 1, 1891, Roll 68, 281.

12. AR, Report of the Commandant of the United States Infantry and Cavalry School, August 1, 1895, Roll 85, 179; Timothy K. Nenninger, *The Leavenworth Schools and the Old Army: Education, Professionalism, and the Officer Corps of the United States Army, 1881–1918* (London and Westport, Conn., 1978), 35.

13. William P. Burnham, "Military Training of the Regular Army of the United States," *JMSIUS* 10 (1889), 623; AR, Report of the Major General Commanding the Army, November 10, 1896, Roll 89, 75; AR, Report of the Department of the Platte, August 28, 1895, Roll 85, 164; "The Condition and Requirements of the Army," *ANR* 18 (August 17, 1895), 98. See also "Practical Field Exercises Under General Forsyth," *ANR* 18 (September 7, 1895), 148.

14. AR, Report of the Inspector General, October 18, 1897, Roll 93, 151–153.

15. Hughes quoted in ibid., October 1, 1890, Roll 64, 107; ibid., September 24, 1894, Roll 81, 92; ibid., September 27, 1895, Roll 95, 113.

16. Alonzo Gray Memoirs, Gray-Woodruff Papers, U.S. Army Military History Institute.

17. Neil C. Mangum, *Battle of the Rosebud: Prelude to the Little Big Horn* (El Segundo, Calif., 1987), 79; unidentified newspaper clipping in the Guy V. Henry Papers, U.S. Army Military History Institute. See also Daniel Appleton to Guy V. Henry, July 13, 1895, Guy V. Henry Correspondence, ibid.

18. AR, Report of First Lt. C. W. Foster, June 25, 1884, in the Report of the Adjutant General, October 15, 1884, Roll 47, 268–269.

19. William T. Sherman to F. V. Greene, October 20, 1879, Roll 45, Letterbooks, Sherman Papers; James S. Pettit, "Tactical Instruction of Officers," *ANJ* 31 (January 20, 1894), 354; James Forance, "Extended Order," *JMSIUS* 17 (July–November 1895), 485; AR, Report of the Department of the Platte, September 1, 1897, Roll 93, 180; Redfield Proctor to John M. Schofield, May 11, 1891, Box 42, Letters Received, John M. Schofield Papers. See also Graham A. Cosmas, *An Army for Empire: The United States Army in the Spanish-American War* (Columbia, Mo., 1971), 42–43.

20. Ambrose, *Upton and the Army*, 140–141; U.S. War Department, *Infantry Drill Regulations, United States Army* (Washington, D.C., 1891), 219; AR, Report of the Infantry and Cavalry School, August 1, 1893, Roll 77, 154.

21. J. H. Reilly, "Some Considerations on the Revision of Our Infantry Tactics," *JMSIUS* 10 (1889), 650–651 (quotations, 651); H. C. Merriam, "The Essentials of Infantry Tactics," *ANR* 12 (April 4, 1891), 213; War Department, *Infantry Drill Regulations*, 186; P. D. Lochridge, "Employment of Cavalry in War," *JUSCA* 6 (1893), 237; "Army Drill Regulations," *ANR* 15 (May 19, 1894), 325.

22. Litchfield, "Military Rifles," 307; AR, Report of the Infantry and Cavalry School, July 11, 1887, Roll 55, 208; Benét quoted in an undated publisher's notice, attached to William H. Morris to John M. Schofield, October 6, 1888, Box 30, Letters Received, John M. Schofield Papers; Henry W. Closson to the Adjutant General, March 31, 1891, Roll 601, Records of the Board on the Revision of Tactics, AGO File 526 AGO 1880, M-689, National Archives. On the new emphasis on the individual soldier, see also "Nomen," "Army Correspondence: Misuse of the Word 'Tactics,'" *ANR* 9 (June 2, 1888), 349, and, for a British Army perspective, "Tactics of the Three Arms," *ANJ* 21 (September 1, 1883), 83.

23. William E. Birkhimer, *Historical Sketch of the Organization, Administration, Matériel, and Tactics of the Artillery, United States Army* (Washington, D.C., 1884), 126, 141; G. N. Whistler, "Artillery Target Practice," *JUSA* 3 (1894), 69; First Lt. Tasker H. Bliss, Report to Maj. Gen. J. M. Schofield, August 1, 1888, Letterbook of Lt. Tasker H. Bliss, Tasker H. Bliss Papers, U.S. Army Military History Institute; AR, Report of the Inspector General of the Army, October 10, 1888, Roll 58, 105.

24. U.S. War Department, *Light Artillery Drill Regulations, United States Army* (Washington, D.C., 1891), 320; U.S. War Department, *Drill Regulations, Light Artillery, United States Army* (Washington, D.C., 1896), 311–312.

25. "The Artillery School," *ANR* 3 (April 29, 1882), 15; W.I.B., "From Fort Monroe," *ANR* 3 (August 12, 1882), 12; Edgar Russel, "Notes on Field Practice," *JUSA* 1 (1892), 106–113. Contemporary journals are rich with articles on artillery target and field practices: G. N. Whistler, "A Few Thoughts on Practical Artillery," *JUSA* 2 (1893), 22–41; Henry C. Davis, "Target Practice," *JUSA* 2 (1893), 42–65; John P. Wisser, trans., "The Artillery-Fire Game," *JUSA* 2 (1893), 122–140, 260–288, 383–407, and 608–639; Charles W. Foster, trans., "Fire-Manoeuvers of Artillery Masses, and the Instruction to be Drawn Therefrom," *JUSA* 2 (1893), 640–657, and *JUSA* 3 (1894), 95–107; Whistler, "Artillery Target Practice," 67–78; John P. Wisser, "The Uses of the Artillery-Fire Game," *JUSA* 4 (1895), 255–264; George F. Landers, "Some Notes on Our Artillery Practice," *JUSA* 7 (January–June 1897), 170–179; C. B. Saterlee, "The Practical Training of Field Batteries," *JMSIUS* 10 (1889), 174–194; and Henry C. Davis, "Light Artillery Target Practice," *JMSIUS* 18 (January–May 1896), 102–112.

26. John P. Wisser, "Practical Instructions in Minor Tactics," *JMSIUS* 8 (1887), 130–158; J. B. Babcock, "Field Exercises and the Necessity for an Authorized Manual of Field Duties," *JMSIUS* 12 (1891), 938–951.

27. Nenninger, *Leavenworth Schools and the Old Army*, 43–47.

28. J. B. Babcock to My Dear Swift, February 2, 1894, Box 23, Letters Re-

ceived, John M. Schofield Papers; W.D.B. [William D. Beach], "A Record of Experience with the Field Sketching Case at the Infantry and Cavalry School," *JUSCA* 8 (1895), 314–316. On the sketch case, see also William D. Beach, "The Practical Work Course in Engineering for Cavalry Officers at the Infantry and Cavalry School," *JUSCA* 9 (1896), 330–335.

29. Guy V. Henry, "Cavalry Instruction," *JUSCA* 7 (1894), 292, 297–299 (quotations 298 and 299); F. S. Foltz, "Things We Are Forgetting," *JMSIUS* 21 (July–November 1897), 284. On the rides of instruction made during this period and later, see also Carol Reardon, *Soldiers and Scholars: The U.S. Army and the Uses of Military History, 1865–1920* (Lawrence, Kans., 1990), 46–47. After the Spanish-American War "staff rides" or "historical rides" of Civil War battlefields became an important part of the army's effort at professional education. Ibid., 50–66.

30. AR, Report of the Artillery School, September 28, 1887, Roll 55, 178; Francis V. Greene, "The Important Improvements in the Art of War During the Past Twenty Years and Their Probable Effect on Future Military Operations," *JMSIUS* 4 (1883), 14; William T. Sherman to E. S. Otis, January 26, 1882, Roll 47, Letterbooks, Sherman Papers; AR, Report of the Infantry and Cavalry School, August 29, 1886, Roll 53, 211; ibid., August 9, 1889, Roll 61, 203. Greene's article was summarized in "Improvements in the Art of War," *ANR* 4 (April 14, 1883), 1–3. On the importance of field entrenchments, see also Beach, "The Practical Work Course," 328–337; James Chester, "Battle Entrenchments and the Psychology of War," *JMSIUS* 7 (1886), 298–312; John G. D. Knight, "The Attack and Defense of Modern Fortifications, and the Latest Experience and Principles in Modern Sieges," *JMSIUS* 8 (1887), 381–404; A. L. Parmerter, "The Use of Field-Works in Military Operations," *JMSIUS* 14 (1893), 1182–1202.

31. David G. Chandler, *The Campaigns of Napoleon* (New York, 1966), 43.

5. The Deadly Ground

1. Sisson C. Pratt, ed., *A Précis of Modern Tactics* (London, 1896), 25; Lt. Col. H. M. Lazelle to the Adjutant General, September 19, 1881, in "Proposed Revision of the Tactics," *ANJ* 19 (November 26, 1881), 359; Francis V. Greene, "The Important Improvements in the Art of War During the Past Twenty Years and Their Probable Effect on Future Military Operations," *JMSIUS* 4 (1883), 24.

2. Lt. Col. H. M. Lazelle to the Adjutant General, September 19, 1881, in "Proposed Revision," 359.

3. Robert M. Utley, *Frontier Regulars: The United States Army and the Indian, 1866–1891* (New York and London, 1973), 70–71 (metallic cartridges); William E. Birkhimer, "Has the Adaptation of the Rifle-Principle to Fire-Arms Diminished the Relative Importance of Field Artillery?" *JMSIUS* 6 (1885), 203 (breech-loading artillery); AR, Report of the Inspector General, October 25, 1889, Roll 61, 136 (smokeless powder).

4. AR, Orders No. 127, Post of Fort Leavenworth, Kansas, May 28, 1884, Roll 47, 192; Junius Brutus Wheeler, *The Elements of Field Fortifications for the Use of the Cadets of the United States Military Academy at West Point, N.Y.* (New York, 1882); AR, Report of the Artillery School, October 1, 1885, Roll 50, 196–197; W. D. Beach to John M. Schofield, September 14, 1894, Box 23, Letters Received, John M. Schofield Papers, Library of Congress.

5. James Chester, "Battle Intrenchments and the Psychology of War," *JMSIUS* 7 (1886), 312; Robert Home, *Précis of Modern Tactics* (London, 1882), 70–71. See also Pratt, ed., *Précis of Modern Tactics*, 25.

6. Wesley Merritt, testimony to the President's Commission on the Conduct of the War with Spain, *Report of the Commission Appointed by the President to Investigate the Conduct of the War Department in the War with Spain*, 8 vols. (Washington, D.C., 1899), 7:3267; James Chester, "Battle Intrenchments," 309, and "Military Misconceptions and Absurdities," *JMSIUS* 14 (1893), 517; H. J. Reilly, "Some Considerations on the Revision of Our Infantry Tactics," *JMSIUS* 10 (1889), 641. See also L. W. V. Kennon, "Considerations Regarding the 'Battle Tactics' of Infantry," *JMSIUS* 7 (1886), 14, and James Chester, "Impending Changes in the Character of War," *JMSIUS* 19 (July–November 1896), 83.

7. "The New Tactics," *ANR* 12 (September 26, 1891), 609; E. S. Godfrey, "Cavalry Fire Discipline," *JMSIUS* 19 (July–November 1896), 255; "Infantry Drill Regulations," *ANR* 21 (April 24, 1897), 268.

8. Grady McWhiney and Perry D. Jamieson, *Attack and Die: Civil War Military Tactics and the Southern Heritage* (University, Ala., 1982), 82–88.

9. AR, Report of Lt. Col. N. B. Sweitzer, August 20, 1884, in the Report of the Adjutant General, October 15, 1884, Roll 47, 236; "Modern Attack Formations," *ANR* 8 (July 30, 1887), 493; "Tactics of the Three Arms," *ANJ* 21 (September 1, 1883), 83.

10. Timothy K. Nenninger, *The Leavenworth Schools and the Old Army: Education, Professionalism, and the Officer Corps of the United States Army, 1881–1918* (London and Westport, Conn., 1978), 43; AR, Report of Lt. Col. N. B. Sweitzer, August 20, 1884, in the Report of the Adjutant General, October 15, 1884, Roll

47, 236; James Chester, "Military Misconceptions and Absurdities," *JMSIUS* 14 (1893), 517; U.S. War Department, *Infantry Drill Regulations, United States Army* (Washington, D.C., 1891), 195, 196.

11. AR, Report of Capt. Theodore Schwan, September 10, 1885, appended to the Report of the Adjutant General, Roll 50, 306; AR, Report of the Department of Texas, August 21, 1894, Roll 81, 146.

12. Bigelow quoted in Russell F. Weigley, *The American Way of War: A History of United States Military Strategy and Policy* (New York and London, 1976), 197.

13. Carol Reardon, *Soldiers and Scholars: The U.S. Army and the Uses of Military History, 1865–1920* (Lawrence, Kans., 1990), 137–141; Edward J. McClernand, "Infantry Battle Tactics," *JMSIUS* 11 (1890), 528; "Machine Guns," *ANR* 8 (March 5, 1887), 149.

14. AR, Report to the Adjutant General of Capt. Edward Field, July 26, 1882, Roll 41, 200; John J. Bigelow, Jr., "The Sabre and Bayonet Question," *JMSIUS* 3 (1882), 66, 95; AR, Report of Capt. William H. Powell, June 30, 1884, Roll 47, 276; C. D. Parkhurst, "The Practical Education of the Soldier," *JMSIUS* 11 (1890), 950–951. See also E. S. Avis, "Practical Work for Infantry," ibid., 420–427.

15. War Department, *Infantry Drill Regulations,* 55–62.

16. James Chester, "Musketry," *JMSIUS* 12 (1891), 232, 239; Matthew F. Steele, "Military Reading: Its Use and Abuse," *JUSCA* 8 (1895), 93; William H. Carter, *From Yorktown to Santiago with the Sixth U.S. Cavalry* (Baltimore, 1900), 279. Historian Russell Gilmore found that the American interest in "field fire" and, more important, the adoption of the Krag–Jorgensen destroyed the consensus in favor of aimed fire over volley fire. During the 1890s, as Gilmore put it, the "Army nearly gave up shooting practice." Russell Gilmore, "'The New Courage': Rifles and Soldier Individualism, 1876–1918," *Military Affairs* 40 (October 1976), 99.

17. AR, Report of the Inspector General of the Army, October 10, 1888, Roll 58, 104; F. S. Strong, "The Necessity for Proper Field Artillery Practice Grounds, and Their Equipment," *JMSIUS* 19 (July–November 1896), 239–240; AR, Report of the Department of the East, August 31, 1894, Roll 81, 103; AR, Report of the Department of the East, September 16, 1896, Roll 89, 132–133 (quotation, 133); AR, Report of First Lt. J. D. C. Hoskins, October 22, 1889, in the Report of the Superintendent of the U.S. Military Academy, September 8, 1890, Roll 64, 237. See also AR, Report of the Superintendent of the U.S. Military Academy, September 1, 1891, Roll 68, 284 and 285, and September 5, 1892, Roll 73, 162 and 163.

18. William T. Sherman to F. V. Greene, January 13, 1878, Roll 45,

Letterbooks, Sherman Papers, Library of Congress. See also AR, Report of the Commanding General of the Army, November 3, 1881, Roll 38, 38.

19. AR, Report of the Inspector General of the Army, October 10, 1888, Roll 58, 104; AR, Report of the U.S. Artillery School, September 15, 1890, Roll 64, 212. See also ibid., September 1, 1892, Roll 73, 150.

20. AR, Report of the U.S. Artillery School, August 31, 1897, Roll 93, 210–211.

21. William E. Birkhimer, "Has the Adaptation of the Rifle-Principle to Fire-Arms Diminished the Relative Importance of Field Artillery?" *JMSIUS* 6 (1885), 223. See also Birkhimer's later essay, "Relative Efficiency of Infantry and Artillery," *JMSIUS* 21 (July–November 1897), 53, 58, 60.

22. William E. Birkhimer, *Historical Sketch of the Organization, Administration, Matériel, and Tactics of the Artillery, United States Army* (Washington, D.C., 1884), 120; AR, Report of the Artillery School, October 17, 1883, Roll 44, 195; AR, Report of the Artillery School, October 1, 1885, Roll 50, 197. See also First Lt. Tasker H. Bliss, Report to Maj. Gen. J. M. Schofield, August 1, 1888, Letterbook of Lt. Tasker H. Bliss, Tasker H. Bliss Papers, U.S. Army Military History Institute.

23. AR, Report of the Inspector General, Military Division of the Atlantic, October 1, 1880, Roll 35, 172; Birkhimer, "Adaptation of the Rifle-Principle," 213.

24. David A. Armstrong, *Bullets and Bureaucrats: The Machine Gun and the United States Army, 1861–1916* (Westport, Conn., and London, 1982), 72, 86–89; Col. E. B. Williston, memorandum to the Board on the Revision of Tactics, 1888, Roll 606, Records of the Board on the Revision of Tactics, AGO File 526 AGO 1880, M-689. Armstrong's book is an excellent study of the army's long effort to define the role and organization of the machine gun. On Williston and the machine gun, see also "Colonel Williston on Machine Guns," *ANR* 7 (May 29, 1886), 338; "Machine Guns," *ANR* 8 (March 5, 1887), 149; *ANR* 8 (December 3, 1887), 771; "Army Correspondence: A Machine Gun Battery," *ANR* 8 (December 17, 1887), 805; Edward B. Williston, "Machine Guns and the Supply of Small-Arm Ammunition on the Battle-Field," *JMSIUS* 7 (1886), 121–160; and John H. Parker, *History of the Gatling Gun Detachment, Fifth Army Corps, at Santiago, with a Few Unvarnished Truths Concerning That Expedition* (Kansas City, Mo., 1898), 92.

25. Armstrong, *Bullets and Bureaucrats*, 89; War Department, *Drill Regulations, Light Artillery*, 422; War Department, *Light Artillery Drill Regulations,* 435; AR, Report of the Department of Dakota, August 25, 1894, Roll 81, 130.

26. AR, Report of the Inspector General, Military Division of the Atlantic,

October 1, 1880, Roll 35, 169; E. L. Zalinski to Dear Bliss, February 28, 1891, Correspondence, Tasker H. Bliss Papers; Charles D. Parkhurst, "Field-Artillery: Its Organization and Its Role," *JUSA* 1 (1892), 273. See also Ernest Hinds, "Light Artillery Target Practice," *JUSA* 4 (1895), 480.

27. John C. Tidball, *Manual of Heavy Artillery Service Prepared for the Use of the Army and Militia of the United States* (Washington, D.C., 1880); John C. Tidball to the Adjutant General, October 31, 1880, Roll 601, Records of the Board on the Revision of Tactics.

28. John C. Tidball to My Dear Caziarc, December 2, 1890, Roll 602, Records of the Board on the Revision of Tactics.

29. Ibid.; War Department, *Light Artillery Drill Regulations,* 393, 399, 400.

30. Vardell Edwards Nesmith, Jr., "The Quiet Paradigm Change: The Evolution of the Field Artillery Doctrine of the United States Army, 1861–1905" (Ph.D. diss., Duke University, 1977), 2–3, 58.

31. Birkhimer, "Adaptation of the Rifle-Principle," 200; War Department, *Light Artillery Drill Regulations,* 424; ibid., *Drill Regulations, Light Artillery,* 411; Joseph E. Kuhn, trans., "A New Method of Indirect Laying for Field Artillery," *JUSA* 3 (1894), 444–474; A. B. Dyer, *Handbook for Light Artillery* (New York, 1896), 365. If the angle was greater than fifteen degrees, Dyer called this "high-angle fire," rather than indirect fire. Ibid. Historian Vardell E. Nesmith, Jr., offered twenty degrees elevation as the general maximum for direct fire. Nesmith, "The Quiet Paradigm Change," 50.

32. War Department, *Light Artillery Drill Regulations,* 406, 428; idem, *Drill Regulations, Light Artillery,* 394, 416. See also Smith, "The Quiet Paradigm Change," 58.

33. War Department, *Light Artillery Drill Regulations,* 422; idem, *Drill Regulations, Light Artillery,* 410.

34. War Department, *Light Artillery Drill Regulations,* 427; idem, *Drill Regulations, Light Artillery,* 414.

35. War Department, *Light Artillery Drill Regulations,* 428; idem, *Drill Regulations, Light Artillery,* 416.

36. War Department, *Light Artillery Drill Regulations,* 423; idem, *Drill Regulations, Light Artillery,* 411.

37. War Department, *Light Artillery Drill Regulations,* 423; idem, *Drill Regulations, Light Artillery,* 410.

38. War Department, *Light Artillery Drill Regulations,* 417; idem, *Drill Regulations, Light Artillery,* 403–404. See also Graham A. Cosmas, "San Juan Hill and

El Caney, 1–2 July 1898," in Charles E. Heller and William A. Stofft, eds., in *America's First Battles*, 1776–1945 (Lawrence, Kans., 1986),115.

39. AR, Report of the Lieutenant General of the Army, November 1, 1884, Roll 47, 51; Wesley Merritt, "Cavalry: Its Organization and Armament," *JMSIUS* 1 (1880), 48; Francis B. Heitman, *Historical Register and Dictionary of the United States Army*, 2 vols. (Washington, D.C., 1903), 1:841; Theodore F. Rodenbough, "Cavalry War Lessons," *JUSCA* 2 (1889), 104. See also Arthur L. Wagner, "Discussions," *JUSCA* 3 (1890), 152.

40. Heitman, *Historical Register*, 1:324; Philip St. George Cooke to William T. Sherman, May 12, 1883, Roll 31, General Correspondence, Sherman Papers; Philip St. George Cooke, *Cavalry Tactics; or Regulations for the Instruction, Formations, and Movements of the Cavalry of the Army and Volunteers of the United States*, 2 pts. (Washington, D.C., 1861); P.S.G.C., "Tactics," *ANJ* 21 (September 15, 1883), 123.

41. George W. Van Deusen, "The Tactical Use of Mounted Troops in Future Campaigns, with Comments on the Recent Rehabilitation of the Lance in European Armies," *JUSCA* 5 (1892), 235; Wesley Merritt, "Cavalry: Its Organization and Armament," *JMSIUS* 1 (1880), 51; Allyn K. Capron, "Some Important Factors in the Instruction of Cavalry," *JUSCA* 10 (1897), 263. A contemporary officer argued the opposing, traditional point of view in "Cavalry in Future Battles," *ANR* 12 (June 27, 1891), 401–402. On the role of cavalry, see also Reardon, *Soldiers and Scholars*, 141–142.

42. S. B. M. Young to the Post Adjutant, Fort Leavenworth, Kansas, June 8, 1883, Journal and Personal Letters, 1874–1883, Samuel B. M. Young Papers, U.S. Army Military History Institute; Bvt. Brig. Gen. Frank Wheaton, July 29, 1881, endorsement to Lt. F. A. Boutelle, July 18, 1881, memorandum, AGO File 5302 AGO 1881, Recommendations of Lt. Col. James W. Forsyth, Roll 94, M–689, National Archives. See also Adjutant General, Military Division of the Pacific and Department of California to the Commanding General, Military Division of the Pacific, May 25, 1881, AGO File 1558 AGO 1882, Recommendations of Lt. Col. James W. Forsyth; and, in the same file: Report of Maj. A. K. Arnold, July 11, 1880; Report of Maj. S. M. Whitside, June 8, 1880; Report of Captain W. A. Rafferty, July 15, 1880; and Report of Capt. E. C. Hentig, August 14, 1880. See also Capt. John A. Kress, July 27, 1881, endorsement to Lt. F. A. Boutelle, July 18, 1881, memorandum, AGO File 5302 AGO 1881, Recommendations of Lt. Col. James W. Forsyth.

43. Lt. Gen. Philip H. Sheridan, April 8, 1882, endorsement to Lt. Col. James

W. Forsyth, March 31, 1882, memorandum, AGO File 1558 AGO 1882, ibid.; Eben Swift, "Sabers or Revolvers?" *JUSCA* 1 (1888), 44.

44. Report of Maj. A. K. Arnold, July 11, 1880, AGO File 1558 AGO 1882, Recommendations of Lt. Col. James W. Forsyth; Van Deusen, "Tactical Use of Mounted Troops," 235; Report of Capt. W. A. Rafferty, July 15, 1880, File 1558 AGO 1882, Recommendations of Lt. Col. James W. Forsyth.

45. E. P. Andrus, "The Saber," *JUSCA* 5 (1892), 378; Moses Harris, "Smokeless Powder in Its Relation to Cavalry Efficiency," *JUSCA* 6 (1893), 40; Peter E. Traub, "On the Saber and Saber Exercise," *JUSCA* 4 (1891), 305; J. P. Ryan, "Some Cavalry Lessons from the Civil War," *JUSCA* 8 (1895), 272.

46. Guy V. Henry to John M. Schofield, September 11, 1888, Box 27, Letters Received, Schofield Papers.

47. Lt. Col. James W. Forsyth to the Adjutant General, Military Division of the Missouri, March 31, 1882, AGO File 1558 AGO 1882, Recommendations of Lt. Col. James W. Forsyth; Col. John C. Kelton, "Observations," August 8, 1881, File 1558 AGO 1882, Recommendations of Lt. Col. James W. Forsyth; Wesley Merritt, "Discussion," *JUSCA* 1 (1888), 36.

48. Heitman, *Historical Register*, 1:859; U.S. War Department, *Cavalry Drill Regulations, United States Army* (Washington, D.C., 1891), 61–88, 361.

49. Heitman, *Historical Register*, 1:284, 447; U.S. War Department, *Drill Regulations for Cavalry, United States Army* (Washington, D.C., 1896), 3, 63–80, 80–90, 361; idem, *Cavalry Drill Regulations,* 62–88.

50. Ryan, "Cavalry Lessons from the Civil War," 282; James Parker, "Cavalry Extended Order Formations," *JUSCA* 7 (1894), 70; War Department, *Cavalry Drill Regulations,* 314–315, 321–325, 328–332; idem, *Drill Regulations for Cavalry,* 314–315, 321–325, 328–332; John Gibbon, "Arms to Fight Indians," *United Service* 1 (April 1874), 243. On dismounted cavalry tactics, see also J. B. Babcock, "Dismounted Service of the Cavalry," *JUSCA* 1 (1888), 88, and an article on this subject by a future World War I general officer, J. T. Dickman, "Value of the Fire of Dismounted Cavalry," *JMSIUS* 16 (January–May 1895), 530–536.

51. AR, Report of the Inspector General of the Army, October 10, 1888, Roll 58, 104; Moses Harris to James W. Forsyth, February 10, 1882, AGO File 1558 AGO 1882, Recommendations of Lt. Col. Forsyth.

52. AR, Report of the Inspector General, November 4, 1896, Roll 89, 108–109.

53. War Department, *Drill Regulations for Cavalry,* 361, 365, 366.

54. Files AGO 1558 AGO 1882 and 5302 AGO 1881, Recommendations of Lt. Col. Forsyth. On mounted firing, see also J. C. Kelton, "Devices by Means of

Which Effective Mounted Firing with the Pistol and Carbine Can Be Obtained by the Cavalry in Attack," *JUSCA* 1 (1888), 60–70; G. B. Sanford, "Mounted Fire Action of Cavalry," *JUSCA* 1 (1888), 79–87; George Paddock, "Mounted Pistol Practice," *JUSCA* 5 (1892), 297–301.

55. General William T. Sherman, September 29, 1881, endorsement to Lt. F. A. Boutelle, July 18, 1881, memorandum, AGO File 5302 AGO 1881, Recommendations of Lt. Col. James W. Forsyth; General Orders No. 57, May 24, 1882, AGO File 1558 AGO 1882, Recommendations of Lt. Col. James W. Forsyth; AR, Report of the Major General Commanding the Army, October 23, 1890, Roll 64, 51.

56. Report of First Lt. T. A. Touey, February 3, 1881; Report of Col. E. A. Carr, February 15, 1881; Report of Capt. E. C. Hentig, August 14, 1880; all in AGO File 1558 AGO 1882, Recommendations of Lt. Col. James W. Forsyth.

6. *Great Changes Now and to Come*

1. Emory Upton, *Infantry Tactics, Double and Single Rank* (New York, 1875, 1876, 1877, 1884, 1885, 1886, 1887, 1889, 1890); U.S. War Department, *Cavalry Tactics, United States Army, Assimilated to the Tactics of Infantry and Artillery* (New York, 1878, 1879, 1883); U.S. War Department, *Artillery Tactics, United States Army, Assimilated to the Tactics of Infantry and Cavalry* (New York, 1876, 1877, 1878, 1882, 1883, 1885, 1886); George Gibson, *Battalion Drill and Battalion Skirmish Drill* (Fort Keough, Mont., 1887), on Roll 605, Records of the Board on the Revision of Tactics, AGO File 526 AGO 1880, M-689, National Archives.

2. AGO, *Official Memoranda of Decisions on Points of Tactics* (Washington, D.C., 1886), Roll 606, Records of the Board on the Revision of Tactics, AGO File 526 AGO 1880, M-689, National Archives; John T. French, Jr., ed., *Interpretations of Infantry Drill Regulations, United States Army* (New York, 1893), foreword, n.p; "The Work of the Tactical Board," *ANJ* 28 (November 8, 1890), 166. For examples of questions and clarifications published in this periodical, see "Tactical Questions," *ANJ* 19 (February 18, 1882), 634; *ANJ* 19 (April 15, 1882), 833; and the many instances throughout *ANJ* 22, passim. For examples in *ANR*, see "Skirmisher," "A Question in Tactics," *ANR* 5 (May 3, 1884), 5; "Common Sense," "Questions in Tactics," *ANR* 5 (November 15, 1884), 12; and "An Important Question in Tactics," *ANR* 7 (March 27, 1886), 204.

3. "The History of Our Tactics," *ANJ* 25 (February 11, 1888), 570; "The Work of the Tactical Board," *ANJ* 28 (November 8, 1890), 166. See also "The United States Army," *ANR* 9 (February 25, 1888), 113.

4. Emory Upton to William T. Sherman, January 30, 1880, Roll 26, General Correspondence, William T. Sherman Papers, Library of Congress; "The Company Column," *ANJ* 12 (May 8, 1875), 616–617. See also Emory Upton to William T. Sherman, February 2, 1880, Sherman Papers; "European and American Tactics," *ANJ* 18 (April 16, 1881), 770–771; Russell F. Weigley, *History of the United States Army* (Bloomington, Ind., 1984), 276–277; Stephen E. Ambrose, *Upton and the Army* (Baton Rouge, 1964), 140–141.

5. Annual Reports of the U.S. War Department, 1822–1907, M-997, National Archives; Report of Lt. Col. N. B. Sweitzer, August 20, 1884, in the Report of the Adjutant General, October 15, 1884, Roll 47, 235–237; ibid., Report of Capt. Edward Field, July 26, 1882, in the Report of the Adjutant General, November 3, 1882, Roll 41, 201; ibid., Report of the Department of the Platte, August 27, 1887, Roll 55, 134.

6. Weigley, *History of the Army*, 273–274; Carol Reardon, *Soldiers and Scholars: The U.S. Army and the Uses of Military History, 1865–1920* (Lawrence, Kans., 1990), 13–15. Reardon skillfully puts the army's use of military history into the larger context of its growing professionalism. On the general subject of professionalism, see also the invaluable Edward M. Coffman, *The Old Army: A Portrait of the American Army in Peacetime, 1784–1898* (New York and Oxford, 1986), 269–286, and, on the Leavenworth Schools in particular, another useful book, Timothy K. Nenninger, *The Leavenworth Schools and the Old Army: Education, Professionalism, and the Officer Corps of the United States Army, 1881–1918* (London and Westport, Conn., 1978).

7. Matthew F. Steele, "Military Reading: Its Use and Abuse," *JUSCA* 8 (1895), 94–95 (quotation, 95). Three helpful sources on the journals and associations are: Michael E. Unsworth, "Army Service Journals, 1879–1917: Vehicles of Professionalism" (paper presented at the fifty-sixth annual meeting of the American Military Institute, April 15, 1989); Weigley, *History of the Army*, 274; and Reardon, *Soldiers and Scholars,* 92–93. On map exercises and war gaming, see J. J. O'Connell, "Kriegspiel of Vinturinus," *JMSIUS* 4 (1883), 418–422; Nenninger, *Leavenworth Schools and the Old Army*, 46–47; and Reardon, *Soldiers and Scholars,* 40–45.

8. C. G. Ayres, "The New Drill Regulations as They Apply to the American Volunteer," *JUSCA* 7 (1894), 303; H. S. Hawkins, "Outline of a Manual of Infantry Drill," *JMSIUS* 11 (1890), 354; Julius A. Penn, "Mounted Infantry," *JMSIUS* 12 (1891), 1148; Charles D. Parkhurst, "Field-Artillery: Its Organization and Its Role," *JUSA* 1 (1892), 263–264 (quotation, 273); "Q.E.D.," "Deliver Us from Our Friends," *ANR* 15 (March 24, 1894), 191. For other primary examples,

see J. B. Babcock, "Field Exercises and the Necessity for an Authorized Manual of Field Duties," *ANR* 15 (March 24, 1894), 950, 951; W. H. Carter, "Tradition and Drill Regulations," *JUSCA* 7 (1894), 114; "Infantry Drill Regulations," *ANR* 21 (April 24, 1897), 269. See also Reardon, *Soldiers and Scholars,* 90, 95–98, 108, 112.

9. Henry W. Closson to the Adjutant General, March 31, 1891, Roll 601, Records of the Board on the Revision of Tactics; "Modern Infantry Fire," *ANJ* 17 (January 3, 1880), 427; "Infantry Fire of the Future," *ANJ* 17 (October 18, 1879), 202–203. See also "Infantry Fire in Modern War," *ANJ* 17 (July 17, 1880), 1033–1034, and "The Tactics of the Future," *ANJ* 33 (July 4, 1896), 801.

10. AR, Report of Lt. Col. N. B. Sweitzer, August 20, 1884, in the Report of the Adjutant General, October 15, 1884, Roll 47, 236.

11. "G.," "Upton's Tactics," *ANJ* 5 (June 13, 1868), 683; "Eugarps," "Upton's Tactics," *ANJ* 5 (May 16, 1868), 618.

12. "Gettysburg," "Upton's Tactics in the Field," *ANJ* 5 (June 6, 1868), 666; "Cultas Wa-waw," *ANJ* 8 (November 19, 1870), 219.

13. Lt. Col. H. M. Lazelle to the Adjutant General, September 19, 1881, in "Proposed Revision of the Tactics," *ANJ* 19 (November 26, 1881), 359; John A. Kress to Dear Friend, January 30, 1880, Roll 605, Records of the Board on the Revision of Tactics. For other criticisms of Upton's tactics, see L. W. V. Kennon, "Considerations Regarding the 'Battle Tactics' of Infantry," *JMSIUS* 7 (1886), 1–45, esp. 27–28; idem, "Some 'Points on Tactics,'" *ANR* 9 (August 25, 1888), 540–541; and H. J. Reilly, "Some Considerations on the Revision of Our Infantry Tactics," *JMSIUS* 10 (1889), 640–660.

14. "Badger," "'Then What Can a Poor Debbil Do?'" *ANJ* 22 (July 4, 1885), 999; William E. Birkhimer, *Historical Sketch of the Organization, Administration, Matériel, and Tactics of the Artillery, United States Army* (Washington, D.C., 1884), vi, 308, 325–330; William H. Powell, "New Tactics Proposed," *ANJ* 25 (March 31, 1888), 712. See also "United States Army," 113.

15. AR, Report of the Inspector General, October 6, 1886, Roll 53, 112; AR, Report of the Lieutenant General of the Army, November 1, 1887, Roll 55, 78; AR, Report of the Major General Commanding the Army, October 23, 1890, Roll 64, 52.

16. William H. Morris, *Field Tactics for Infantry* (New York, 1864); William H. Morris, *Infantry Tactics* (New York, 1865); William H. Morris, *Tactics for Infantry Armed with Breech-Loading or Magazine Rifles* (New York, 1882), Roll 598, and William H. Morris, "Tactics for Infantry Armed with Breech-Loading or Magazine Rifles," 1888, Roll 605, Records of the Board on the Revision of Tac-

tics; William H. Morris to John M. Schofield, October 6, 1888, Box 30, Letters Received, John M. Schofield Papers, Library of Congress.

17. G. N. Whistler, "A Manual of Drill and Tactics," *JMSIUS* 4 (1883), 55–66; John H. Patterson, *Infantry Tactics; A System of Attack and Defense With Reenforced Skirmish Lines, for the Instruction of Officers and Non-Commissioned Officers; School of the Company and Battalion* (Fort Leavenworth, 1884), 1, Roll 606, Records of the Board on the Revision of Tactics.

18. Robert M. Utley, *Custer and the Great Controversy: The Origin and Development of a Legend* (Pasadena, Calif., 1980), 67.

19. Correspondence on the Proposal of Captain R. P. Hughes, Roll 254, AGO File 310 AGO 1884, M-689, National Archives.

20. H. S. Hawkins, "A New System of Tactics," *ANJ* 24 (June 11, 1887), 913, 919. Hawkins gave something of a "sour grapes" account of the fate of his proposal in "Outline of a Manual of Infantry Drill," *JMSIUS* 11 (1890), 353–354. See also "The United States Army," *ANR* 10 (April 6, 1889), 209.

21. "Livermore's Manoeuvers for Infantry," *ANJ* 25 (February 18, 1888), 592–593. Livermore's work was published, but not authorized by the War Department. See "Manoeuvers for Infantry," *ANJ* 25 (May 12, 1888), 838–839 for a favorable review of the work.

22. Frank H. Edmunds, "Battle Tactics," *JMSIUS* 12 (1891), 1202–1210.

23. "Revised Upton," *ANR* 8 (October 29, 1887), 689.

24. When a board of officers began meeting in March 1888 to prepare a new tactics, it found itself in "almost daily receipt of . . . various ideas" and proposals. "The Board on Tactics," *ANJ* 25 (March 10, 1888), 659. See also Charles W. Hobbs, "Suggestions on the Subject of Infantry Organization and Tactics," *JMSIUS* 6 (1885), 252–260.

25. Special Orders No. 14, January 18, 1888, Roll 601, Records of the Board on the Revision of Tactics; Francis B. Heitman, *Historical Register and Dictionary of the United States Army*, 2 vols. (Washington, D.C., 1903), 1:199.

26. Special Orders No. 14, January 14, 1888, and memoranda of the Bates Board to the Adjutant General, December 1, 1890, Roll 601, Records of the Board on the Revision of Tactics; Heitman, *Historical Register*, 1:461.

27. "The History of Our Tactics," *ANJ* 25 (February 11, 1888), 570; "The Board on Tactics," *ANJ* 25 (March 10, 1888), 659.

28. Special Orders No. 68, March 23, 1889, Roll 602, Records of the Board on the Revision of Tactics; memorandum of the Board on the Revision of Tactics to the Adjutant General, December 1, 1890, Roll 601, Records of the Board on the Revision of Tactics; "The Tactical Board," *ANJ* 26 (April 20, 1889), 685. See

also "New Drill Regulations for the Cavalry," *JUSCA* 3 (1890), 286; "The United States Army," *ANR* 10 (April 6, 1889), 209; *ANR* 10 (July 27, 1889), 365; and *ANR* 11 (October 4, 1890), 633.

29. Lt. Col. J. C. Bates to the Adjutant General, December 2, 1890, and January 15, 1891, Roll 601, Records of the Board on the Revision of Tactics.

30. Wesley Merritt to John M. Schofield, March 17, 1891, Henry W. Closson to the Adjutant General, March 31, 1891, Thomas H. Ruger to the Adjutant General, July 1, 1891, and John M. Schofield to the Secretary of War, September 21, 1891, Roll 601, Records of the Board on the Revision of Tactics; U.S. War Department, *Infantry Drill Regulations, United States Army* (New York, 1891), 2. On the "staffing" of the Leavenworth tactics, see also "The United States Army," *ANR* 12 (January 10, 1891), 17, and "The New Tactics," *ANR* 12 (September 26, 1891), 609.

31. On Redfield Proctor as an activist secretary of war, see Coffman, *The Old Army*, 224, 233, 259, 281, 332, 375, 377, 378, 397, and Chester W. Bowie, "Redfield Proctor: A Biography" (Ph.D. diss., University of Wisconsin-Madison, 1980), 162–272. Bowie concluded that Proctor, "an outspoken advocate of progressive policies and an effective implementer of sorely needed reforms," was "the most talented" secretary to serve between Edwin Stanton and Elihu Root. Ibid., 271 and 272.

32. War Department, *Infantry Drill Regulations*, 2.

33. "Extended Order Drill," *ANR* 14 (May 20, 1893), 319; Special Orders No. 14, January 14, 1888, Roll 601, Records of the Board on the Revision of Tactics; J. T. French, Jr., "Extended Order Drill," *ANR* 20 (July 11, 1896), 18–19; J. T. French, Jr., *Revision of Interpretations of Infantry Drill Regulations* (New York, 1893).

34. C. J. Crane, "Our New Drill Regulations for the Infantry," *JMSIUS* 13 (1892), 1148; C. G. Ayres, "The New Drill Regulations as They Apply to the American Volunteer," *JUSCA* 7 (1894), 301. One correspondent of the *Army and Navy Journal* referred to French as the successor of Scott, Hardee, Casey, and Upton, but even this soldier called the new manual the "Drill Regulations" rather than "French's Tactics." "Verbum Sat Sapienti," "'Drill Regulations'—Present and Future," *ANJ* 32 (October 6, 1894), 92.

35. W. Merritt to John M. Schofield, June 7, 1889, Box 41, Special Correspondence, John M. Schofield Papers; "Drill Regulations for Cavalry, United States Army," *JUSCA* 2 (1889), 173–207, 269–322, 378–409; *JUSCA* 3 (1890), 57–98, 164–203, 286–339; *JUSCA* 4 (1891), 69–84.

36. "Drill Regulations for Infantry, U.S. Army," *ANR* 11 (August 2, 1890),

490–494; *ANR* 11 (August 16, 1890), 523–525; *ANR* 11 (August 23, 1890), 539–540; *ANR* 11 (September 6, 1890), 572–573; *ANR* 11 (September 13, 1890), 588; *ANR* 11 (September 20, 1890), 604; *ANR* 11 (September 27, 1890), 619–620; *ANR* 11 (October 4, 1890), 636; *ANR* 11 (October 11, 1890), 652; *ANR* 11 (October 18, 1890), 668–669; *ANR* 11 (October 25, 1890), 682; *ANR* 11 (November 1, 1890), 700.

37. U.S. War Department, *Infantry Drill Regulations, United States Army* (Washington, D.C., 1891); Hugh T. Reed, *Abridgement of the Drill Regulations for Infantry* (Chicago, 1891), and correspondence relating to this work, Roll 602, Records of the Board on the Revision of Tactics; U.S. War Department, *Infantry Drill Regulations, United States Army* (New York, 1891, 1892, 1893, 1895).

38. U.S. War Department, *Light Artillery Drill Regulations, United States Army* (Washington, D.C., 1891); U.S. War Department, *Cavalry Drill Regulations, United States Army* (Washington, D.C., 1891).

39. Grady McWhiney and Perry D. Jamieson, *Attack and Die: Civil War Military Tactics and the Southern Heritage* (University, Ala., 1982), 67; Paddy Griffith, *Battle Tactics of the Civil War* (New Haven, Conn., 1989), 114–115; John P. Wisser, "Practical Instructions in Minor Tactics," *JMSIUS* 8 (1887), 130; Emory Upton to William T. Sherman, January 30, 1880, Roll 26, General Correspondence, Sherman Papers. See also "Nomen," "Army Correspondence: Misuse of the Word 'Tactics,'" *ANR* 9 (June 2, 1888), 349.

40. L. W. V. Kennon, "Some 'Points on Tactics,'" *ANR* 9 (August 25, 1888), 540–541 (quotation, 541); William P. Burnham, "Military Training of the Regular Army of the United States," *JMSIUS* 10 (1889), 619–620; War Department, *Infantry Drill Regulations,* 10–176, 186–230.

41. War Department, *Infantry Drill Regulations*, 63, 171. Files represented the columns of a company that extended from the unit's front to its rear, as opposed to ranks, the rows of men across its front.

42. Ibid., 8, 63, 65–66.

43. Ibid., 63, 66–67. On the use of Upton's fours in the 1891 tactics, see also L. W. V. Kennon, "The Tactical Board and Its Critics," *ANR* 11 (December 27, 1890), 828–829.

44. War Department, *Infantry Drill Regulations,* 186.

45. Ibid.; "The New Tactics," *ANR* 12 (September 26, 1891), 609; "The New Drill Regulations," *ANR* 11 (October 18, 1890), 665.

46. War Department, *Infantry Drill Regulations,* 186, 194.

47. Ibid., 195, 196; Charles Johnson Post, *The Little War of Private Post* (Boston and Toronto, 1960), 22, and see also 64.

48. War Department, *Infantry Drill Regulations*, 196; idem, *Cavalry Drill Regulations*, 328–331.

49. J. F. C. Fuller, *A Military History of the Western World*, 3 vols. (New York, 1954–1956); Archer Jones, *The Art of War in the Western World* (Urbana, Ill., and Chicago, 1987).

50. McWhiney and Jamieson, *Attack and Die*; Edward Hagerman, *The American Civil War and the Origins of Modern Warfare: Ideas, Organization, and Field Command* (Bloomington, Ind., 1988); Griffith, *Battle Tactics*.

51. War Department, *Light Artillery Drill Regulations*, 408–411; idem, *Cavalry Drill Regulations*, 361–365, 368–369; idem, *Infantry Drill Regulations*, 207–215.

52. War Department, *Infantry Drill Regulations*, 219–226.

53. Ibid., 207, 208, 225.

54. Ibid., 63, 64.

55. Ibid., 63, 186, 194–196. See also "The New Drill Regulations," 665.

56. AR, Report of the Secretary of War, November 24, 1896, Roll 89, 7; Emory Upton, *Infantry Tactics*, 149. The single-battalion regiment was the standard organization of Civil War volunteer units. A regular regiment might be organized into two or more battalions, as for example were both the Twelfth and Fourteenth Infantry regiments during the Antietam campaign. U.S. War Department, *The War of the Rebellion: A Compilation of the Official Records of the Union and Confederate Armies*, 128 vols. (Washington, D.C., 1880–1901), ser. 1, 19, pt. 1:175.

57. AR, Report of the Secretary of War, November 26, 1894, Roll 81, 5–6; "European and American Tactics," *ANJ* 18 (April 16, 1881), 770–771; Ambrose, *Upton and the Army*, 140–141. See also L. W. V. Kennon, "'Battle Tactics' of Infantry," 42–45.

58. AR, Report of the Secretary of War, November 26, 1894, Roll 81, 6.

59. "European and American Tactics," 770–771; Ambrose, *Upton and the Army*, 140–141.

60. AR, Report of the Secretary of War, November 23, 1889, Roll 61, 5; AR, Report of the Secretary of War, November 3, 1891, Roll 68, 12.

61. AR, Report of the Major General Commanding the Army, October 23, 1890, Roll 64, 46; AR, Report of the Major General Commanding the Army, September 24, 1891, Roll 68, 57; September 30, 1892, Roll 73, 47–48; October 1, 1894, Roll 81, 64.

62. AR, Report of the Major General Commanding the Army, November 5, 1895, Roll 85, 68. Miles later shifted his position. AR, Report of the Major General Commanding the Army, November 10, 1896, Roll 89, 78.

63. AR, Report of the Secretary of War, November 23, 1889, Roll 61, 5; AR, Report of the Secretary of War, November 15, 1890, Roll 64, 11–12; AR, Report of the Secretary of War, November 3, 1891, Roll 68, 12.

64. AR, Report of the Secretary of War, n.d., Roll 73, 3.

65. AR, Report of the Secretary of War, November 24, 1894, Roll 81, 5–6; ibid., November 26, 1895, Roll 85, 7; ibid., November 24, 1896, Roll 89, 7.

66. AR, Report of the Secretary of War, November 24, 1894, Roll 81, 5. If the three-battalion organization had been adopted that year, it would also have required adding two foot batteries to each of the army's five artillery regiments. Ibid.

67. Weigley, *History of the Army*, 295–296; War Department, *Infantry Drill Regulations*, 94, 139. On the three-battalion regiment, see also Curtis V. Hard, *Banners in the Air: The Eighth Ohio Volunteers and the Spanish-American War*, ed. Robert H. Ferrell (Kent, Ohio, and London, 1988), 5; Thomas M. Anderson, "Comment and Criticism: Infantry Battle Tactics," *JMSIUS* 11 (1890), 652; F. H. Edmunds, "Is the Three Battalion Organization the Best One for Us?" *JMSIUS* 14 (1893), 749–754; and H. R. Brinkerhoff, "Popular Impressions of the Army," *ANR* 18 (September 7, 1895), 148–149.

68. Richard I. Wolf, "Arms and Innovation: The United States Army and the Repeating Rifle" (Ph.D. diss., Boston University, 1981); AR, Report of the Secretary of War, November 3, 1891, Roll 68, 9. See also "Our New Service Rifle," *ANR* 13 (September 10, 1892), 588–589.

69. AR, Report of the Secretary of War, November 27, 1893, Roll 77, 12; ibid., November 26, 1894, Roll 81, 14; U.S. War Department, *Infantry Drill Regulations: The Manual of Arms, Adapted to the Magazine Rifle, Caliber .30* (Washington, D.C., 1897); "Firing Regulations for the Magazine Rifle, Caliber .30," *ANR* 19 (May 9, 1896), 290.

70. "The New Drill Regulations," *ANR* 11 (October 11, 1890), 649; *ANR* 11 (October 18, 1890), 665; Kennon, "Tactical Board and Its Critics," 829. In 1894, when the Leavenworth tactics had come in for some criticism, the *Army and Navy Register* stood by this initial endorsement. "Army Drill Regulations," *ANR* 15 (May 19, 1894), 325.

71. W. H. Carter, "Tradition and Drill Regulations," *JUSCA* 7 (1894), 116; Capt. F. K. Ward, "Notes on the Cavalry Drill Regulations," January 4, 1893, Roll 604, Records of the Board on the Revision of Tactics.

72. Capt. S. M. Mills to Col. H. C. Hasbrouck, January 28, 1889, Roll 606, Records of the Board on the Revision of Tactics; War Department, *Light Artillery Drill Regulations*, 60–64, 64–79, 397–436.

73. War Department, *Infantry Drill Regulations,* 3–6, 239–240 (quotations, 6 and 239–240).

74. AR, Report of the Artillery School, September 13, 1886, Roll 53, 196; AR, Report of the Artillery School, September 28, 1887, Roll 55, 189; AR, Report of the Infantry and Cavalry School, August 1, 1893, Roll 77, 153. See also C. B. Mayne, *Infantry Fire Tactics* (Chatham, England, 1885).

75. AR, Report of the Infantry and Cavalry School, August 1, 1893, Roll 77, 152; AR, Report of the Infantry and Cavalry School, August 1, 1894, Roll 81, 169; Nenninger, *Leavenworth Schools and the Old Army,* 39–43.

7. No Final Tactics

1. Thomas H. Ruger to the Adjutant General, July 1, 1891, Roll 601, Records of the Board on the Revision of Tactics, AGO File 526 AGO 1880, M-689, National Archives; "The Work of the Tactical Board," *ANJ* 28 (November 8, 1890), 166.

2. "Verbum Sat Sapienti," "'Drill Regulations'—Present and Future," *ANJ* 32 (October 6, 1894), 91–92; C. J. Ayres, "The New Drill Regulations as They Apply to the American Volunteer," *JUSCA* 7 (1894), 301, 303.

3. T. M. Anderson, "Comment and Criticism: Drill Regulations," *JMSIUS* 12 (1891), 150–153 (quotation, 150); Henry M. Lazelle, "The Work of the Tactical Board," *ANJ* 28 (November 8, 1890), 166; C. J. Crane, "Our New Drill Regulations for the Infantry," *JMSIUS* 13 (1892), 1172; AR, Report of the Inspector General, September 28, 1893, Roll 80, 4:719.

4. "Army Drill Regulations,"*ANR* 15 (May 19, 1894), 325; "General Ruger Hard at Work," *ANR* 17 (June 8, 1895), 377; "Infantry Drill Regulations," *ANR* 21 (February 6, 1897), 91; AR, Report of the Inspector General, September 28, 1893, Roll 80, 4:719.

5. D. C. Kingman, "'The Right (or Left) Turn' of the Infantry Drill Regulations," *JMSIUS* 16 (January–May 1895), 537–548; E. C. Brooks, "A Technical Criticism of Our Infantry Drill Book," *JMSIUS* 17 (July–November 1895), 97–107; P. Borger, "The Squad Formation: A Few Suggestions for Its Improvement," *JMSIUS* 17 (January–May 1896), 97–102.

6. Lt. D. L. Brainard, "Drill Regulations," March 13, 1893, Roll 605, Records of the Board on the Revision of Tactics; James Parker, "Cavalry Extended Order Formations," *JUSCA* 7 (1894), 62; W. H. Smith, "Some Remarks on Our New

Cavalry Drill Regulations," *JMSIUS* 14 (1893), 525–539; Ernest Hinds, "Light Artillery Target Practice," *JUSA* 4 (1895), 480.

7. "Army Drill Regulations," *ANR* 15 (May 19, 1894), 325; untitled article in *ANR* 16 (December 29, 1894), 406.

8. "Army Drill Regulations," *ANR* 19 (May 30, 1896), 344; *ANR* 18 (July 13, 1895), 25.

9. U.S. War Department, *Drill Regulations, Light Artillery, United States Army* (Washington, D.C., 1896), 3; U.S. War Department, *Drill Regulations for Cavalry, United States Army* (Washington, D.C., 1896), 3; "General Ruger Hard at Work," *ANR* 17 (June 8, 1895), 377; "Army Drill Regulations," *ANR* 19 (May 30, 1896), 344; "Infantry Drill Regulations," *ANR* 21 (January 9, 1897), 25.

10. War Department, *Drill Regulations, Light Artillery*, 3; untitled article, *ANR* 16 (December 29, 1894), 406.

11. War Department, *Drill Regulations, Light Artillery*, 3.

12. U.S. War Department, *Light Artillery Drill Regulations, United States Army* (Washington, D.C., 1891), 27; War Department, *Drill Regulations, Light Artillery*, 30.

13. War Department, *Light Artillery Drill Regulations*, 80–94; idem, *Drill Regulations, Light Artillery*, 84–86.

14. War Department, *Drill Regulations for Cavalry*, 3; untitled article in *ANR* 16 (December 29, 1894), 406.

15. War Department, *Drill Regulations for Cavalry*, 3.

16. U.S. War Department, *Cavalry Drill Regulations, United States Army* (Washington, D.C., 1891), 11; idem, *Drill Regulations for Cavalry*, 13–14.

17. War Department, *Cavalry Drill Regulations*, 314–315; ibid., *Drill Regulations for Cavalry*, 203–204.

18. "Cavalry Drill Regulations," *ANR* 20 (December 12, 1896), 383.

19. "General Ruger Hard at Work," *ANR* 17 (June 8, 1895), 377; untitled article, *ANR* 18 (August 17, 1895), 98; "Army Drill Regulations," *ANR* 19 (May 30, 1896), 345.

20. "Infantry Drill Regulations," *ANR* 21 (January 23, 1897), 56–57; *ANR* 21 (February 6, 1897), 91; "Quien Sabe?" "The Promised Drill Regulations," *ANR* 21 (April 3, 1897), 224.

21. "Infantry Drill Regulations," *ANR* 21 (April 3, 1897), 220–221; AR, Report of the Commanding General of the Army, November 5, 1895, Roll 85, 68; AR, Report of the Commanding General of the Army, November 10, 1896, Roll 89, 78; "Infantry Drill Regulations," *ANR* 21 (April 24, 1897), 268.

22. Swift quoted in Carol Reardon, *Soldiers and Scholars: The U.S. Army and the Uses of Military History, 1865–1920* (Lawrence, Kans., 1990), 49; Capt. George H. Palmer, "Infantry Tactics," Roll 605, Records of the Board on the Revision of Tactics; William N. Blow, Jr., "'Extended Order' and 'Skirmish Firing' Assimilated," *JMSIUS* 15 (1894), 111. On the army's lack of a broadly accepted doctrine, see Reardon, *Soldiers and Scholars,* 144. On the army's gradual process of professionalization, see Timothy K. Nenninger, *Education, Professionalism, and the Officer Corps of the United States Army, 1881–1918* (London and Westport, Conn., 1978), 6.

23. William H. Morris, *Tactics for Infantry Armed with Breech-Loading or Magazine Rifles* (New York, 1882), Roll 587, Records of the Board on the Revision of Tactics, AGO File 526 AGO File 1880, M-689, National Archives; Charles W. Hobbs, "Suggestions on the Subject of Infantry Organization and Tactics," *JMSIUS* 6 (1885), 257; Arthur L. Wagner, *The Campaign of Königgrätz; A Study of the Austro-Prussian Conflict in Light of the American Civil War* (Kansas City, Mo., 1899), 104–105. See also Nenninger, *Leavenworth Schools and the Old Army,* 41.

24. T. M. Anderson, "Comment and Criticism: Drill Regulations," *JMSIUS* 12 (1891), 151; W. V. Richards, "Is the Tendency of Modern Drill Regulations Salutary?" *JMSIUS* 13 (1892), 905–906 (quotation, 906); Maj. Joseph P. Sanger quoted in AR, Report of the Inspector General, October 18, 1897, Roll 93, 125; AR, Report of the Inspector General, October 18, 1897, Roll 93, 126; Charles Johnson Post, *The Little War of Private Post* (Boston and Toronto, 1960), 22. This opinionated private ignored a point made by Brigadier General Frederick Funston: a regiment had to learn some of the simpler, close-order drill movements before it could train for extended-order fighting. Frederick Funston, *Memories of Two Wars: Cuban and Philippine Experiences* (New York, 1911), 162.

25. James Chester, "Battle Intrenchments and the Psychology of War," *JMSIUS* 7 (1886), 309, and "Impending Changes in the Character of War," *JMSIUS* 19 (July–November 1896), 84–85; "Badger," "'Then What Can a Poor Debbil Do?'" *ANJ* 22 (July 4, 1885), 999; H., "Ignorance of Tactics in the Army," *ANJ* 22 (January 3, 1885), 451. On James Chester, see Carol Reardon, *Soldiers and Scholars,* 24.

26. Capt. John A. Kress, July 27, 1881, endorsement to Lt. F. A. Boutelle, July 18, 1881, memorandum, AGO File 5302 AGO 1881, Recommendations of Lt. Col. James W. Forsyth, Roll 94, M-689, National Archives; AR, Report of the Lieutenant General Commanding the Army, November 1, 1884, Roll 47, 49; "Remarks of General Sheridan at the Philadelphia Banquet," *ANR* 8 (September 24, 1887),

613; Lt. Theodore B. M. Mason to My dear Colonel, November 11, 1884, Box 35, General Correspondence, Philip H. Sheridan Papers, Library of Congress.

27. AR, Report of the Secretary of War, November 30, 1886, Roll 53, 17; A. M. D. McCook to William T. Sherman, May 2, 1882, Roll 30, General Correspondence, Sherman Papers, Library of Congress; Matthew F. Steele, "Military Reading: Its Use and Abuse," *JUSCA* 8 (1895), 103; AR, Report of the Headquarters U.S. Infantry and Cavalry School, October 8, 1884, Roll 47, 152–153; Lt. D. L. Brainard, "Drill Regulations," March 13, 1893, Roll 605, Records of the Board on the Revision of Tactics; AR, Report of the Division of the Pacific, September 17, 1886, Roll 53, 139 and 140.

28. E. V. Sumner, "The Individual Soldier," *JUSCA* 1 (1888), 188; Guy V. Henry, comment on Albert G. Brackett, "Our Cavalry: Its Duties, Hardships, and Necessities, at Our Frontier Posts," *JMSIUS* 4 (1883), 397. See also Sumner's comment on Brackett, ibid., 395–396.

29. W. H. Carter, "Tradition and Drill Regulations," *JUSCA* 7 (1894), 115.

30. AR, Report of the Department of the Missouri, September 22, 1881, Roll 38, 125; Robert Wooster, *The Military and United States Indian Policy, 1865–1903* (London and New Haven, Conn., 1988), 214.

31. Russell F. Weigley, *History of the United States Army* (Bloomington, Ind., 1984), 598; AR, Returns B and F, Report of the Adjutant General, October [n.d.] 1883, Roll 44, table facing page 58-2-W, and 73–78; AR, Report of the Lieutenant General of the Army, November 1, 1887, Roll 55, 78–79; AR, Report of the Division of the Atlantic, September 20, 1887, 114; AR, Report of the Department of the Columbia, August 25, 1896, Roll 89, 156.

32. John R. Brooke to John M. Schofield, March 26, 1890, Box 22, Letters Received, John M. Schofield Papers, Library of Congress; AR, Report of the Major General Commanding the Army, October 23, 1890, Roll 64, 51; AR, Report of the Department of the East, September 16, 1896, Roll 89, 132; ibid., Report of the Inspector General, September 20, 1894, Roll 81, 92. See also "A.P., Second Infantry," "Army Correspondence: Practical Instruction in Campaigning, Etc.," *ANR* 8 (October 15, 1887), 669.

33. Capt. F. K. Ward, "Notes on the Cavalry Drill Regulations," January 4, 1893, Roll 604, Records of the Board on the Revision of Tactics; C. A. P. Hatfield, "The Tendency of Evolution in the Army," *JMSIUS* 21 (July–November 1897), 448; Allyn K. Capron, "Some Important Factors in the Instruction of Cavalry," *JUSCA* 10 (1897), 260.

34. Tasker H. Bliss to John M. Schofield, October 17, [1888], Letterbook of

Letters Sent, August 19, 1888, to July 6, 1889, Tasker H. Bliss Papers, U.S. Army Military History Institute; AR, Report of the Infantry and Cavalry School, October 6, 1883, Roll 44, 201–202; John P. Wisser, "Practical Instruction of Officers at Posts," *JMSIUS* 9 (1888), 199.

35. AR, Report of the Department of the Platte, September 1, 1897, Roll 93, 179; Frederic Remington, letter to the *Sun*, May 28, 1893, newspaper clipping in File R, 1893, Box 33, Letters Received, John M. Schofield Papers. For a reply to Remington, see "Not a Very Old Fogy," "Attack on the Army Refuted," *ANR* 14 (June 17, 1893), 382.

36. E. V. Sumner to Philip H. Sheridan, January 31, 1883, Box 34, General Correspondence, Philip H. Sheridan Papers; E. V. Sumner, "American Practice and Foreign Theory," *JUSCA* 3 (1890), 150; Powhatan H. Clarke, "Our Army All Wrong," *ANR* 14 (May 13, 1893), 302; George W. Cullum, *Biographical Register of the Officers and Graduates of the U.S. Military Academy at West Point, New York*, 9 vols. (Cambridge, Mass., 1901), 4:398; Henry L. Harris to My dear Bliss, September 25, 1889, Correspondence, Tasker H. Bliss Papers.

37. F. S. Strong, "The Necessity for Proper Field Artillery Practice Grounds, and Their Equipment," *JMSIUS* 19 (July–November 1896), 240; AR, Report of the Department of the East, August 31, 1894, Roll 81, 104; AR, Report of the Inspector General, September 27, 1895, Roll 85, 113; AR, Report of the Department of the Platte, September 1, 1897, Roll 93, 179.

38. Wisser, "Practical Instruction," 199; Capron, "Some Important Factors," 261; J. B. Babcock to My Dear Swift, February 22, 1894, Box 23, Letters Received, John M. Schofield Papers; AR, Report of the Department of the Platte, August 28, 1895, Roll 85, 164; "Field Exercises of the Third Cavalry," *ANR* 22 (September 25, 1897), 211; James Chester, "Impending Changes in the Character of War," *JMSIUS* 19 (July–November 1896), 85.

39. Post, *Little War*, 63; Don Russell, *Campaigning with King: Charles King, Chronicler of the Old Army*, ed. Paul Hedren (London and Lincoln, Nebr., 1991), 112; AR, Report of Capt. William H. Powell, June 30, 1884, Roll 47, 273; AR, Report of Capt. George B. Rodney, September 19, 1882, Roll 41, 205; AR, Report of Maj. Jared A. Smith and Lt. Edward L. Randall, September 28, 1882, Roll 41, 224 and 225. On the uneven quality of early National Guard drills, see also John K. Mahon, *History of the Militia and National Guard* (New York and London, 1983), 114, and Graham A. Cosmas, *An Army for Empire: The United States Army in the Spanish-American War* (Columbia, Mo., 1971), 13.

40. Unidentified newspaper clipping in the Guy V. Henry Papers, U.S. Army

Military History Institute; John C. Gresham, "The School at Fort Riley," *JMSIUS* 18 (January–May 1896), 510; A. K. Arnold, "A Field Exercise at Fort Riley," *JUSCA* 10 (1897), 168; G. L. W. Grierson, "Rules for Umpires at Peace Maneuvers, from German Field Service Regulations," *JUSCA* 3 (1890), 107.

41. Quoted in J. T. Dickman, "A Field Exercise at Fort Leavenworth," *JUSCA* 10 (1897), 160. A similar rule prevailed at Fort Riley's Cavalry and Light Artillery School. A. K. Arnold, "A Field Exercise at Fort Riley," *JUSCA* 10 (1897), 169.

42. *Army and Navy Gazette* article, September 19, 1891, quoted in *JUSCA* 4 (1891), 319; AR, Report of the Cavalry and Light Artillery School, December 20, 1893, Roll 81, 163.

43. AR, Report of the Cavalry and Light Artillery School, 164.

44. Wilton C. Hall to John M. Schofield, February 5, 1895, Box 28, Letters Received, Schofield Papers; F. G. Chapman to John M. Schofield, August 12, 1892, Box 24, Letters Received, Schofield Papers; Jackson Kirkman to Wade Hampton, December [n.d.] 1889, Box 29, Letters Received, Schofield Papers; William D. Riley to John M. Schofield, November 13, 1891, Box 32, Letters Received, Schofield Papers. It was during this period, too, that the army experimented with the bicycle for troop transportation, field exercises, and courier service. AR, Report of the Department of the Missouri, September 14, 1892, Roll 73, 104; AR, Report of the Commandant of the Artillery School, August 31, 1896, Roll 89, 183; "The Bicycle in the Field," *ANR* 18 (November 23, 1895), 353; "Army Bicycles," *ANR* 19 (February 1, 1896), 70; and "The Bicycle for Military Purposes," *ANR* 20 (August 29, 1896), 130–131.

45. Major General Frederick Funston's memoirs describe the Sims-Dudley dynamite gun and its first use in combat; see Funston, *Memories of Two Wars*, 119. On the dynamite gun during the Spanish-American War, see AR, Report of Sgt. Hallett Alsop Borrowe, July 14, 1898, 354; AR, Report of Brig. Gen. Leonard Wood, July 14, 1898, 354–355; and David F. Trask, *The War with Spain in 1898* (New York and London, 1981), 362.

46. There are many sources, contemporary and historical, on the improvements in weapons during the late nineteenth century. See, for example, William E. Birkhimer, *Historical Sketch of the Organization, Administration, Matériel, and Tactics of the Artillery, United States Army* (Washington, D.C., 1884); Konrad F. Schrier, Jr., "U.S. Army Field Artillery Weapons, 1866–1918," *Military Collector and Historian* 20 (Summer 1968), 40–45; Richard I. Wolf, "Arms and Innovation: The United States Army and the Repeating Rifle" (Ph.D. diss., Boston University, 1981); Sidney B. Brinckerhoff and Pierce Chamberlain, "The Army's Search

for a Repeating Rifle, 1873–1903," *Military Affairs* 32 (April 1968), 20–30; Charles B. Norton, *American Inventions and Improvements . . .* (Springfield, Mass., 1880); George W. Morse, "Army Correspondence: George W. Morse's Military Inventions," *ANR* 9 (February 11, 1888), 92; "The New Smokeless Powder," *ANR* 10 (July 13, 1889), 433; Joseph Strauss, "Smokeless Powder," *U.S. Naval Institute Proceedings* 27 (December 1901), 733–738; Laurence V. Benet, "A Study of the Effects of Smokeless Powder in a 57mm Gun," *JUSA* 1 (1892), 207–223; E. L. Zalinski, "Comment and Criticism," *JUSA* 1 (1892), 411–412; C. D. Parkhurst, "Electricity and the Art of War," *JUSA* 1 (1892), 315–363, and *JUSA* 2 (1893), 95–121; Henry C. Davis, trans., "The Importance of Smokeless Powder in War," *JUSA* 3 (1894), 108–122; Henry M. Lazelle, "Important Improvements in the Art of War During the Past Twenty Years and Their Probable Effect on Future Military Operations," *JMSIUS* 3 (1882), 307–373 .

47. Francis V. Greene, "The Important Improvements in the Art of War During the Past Twenty Years and Their Probable Effect on Future Military Operations," *JMSIUS* 4 (1883), 22–23; "European and American Tactics," *ANJ* 18 (April 16, 1881), 771.

48. Henry W. Closson to the Adjutant General, March 31, 1891, Roll 601, Records of the Board on the Revision of Tactics; Greene, "The Important Improvements," 41; "Forward," "Thoughts Pertinent to the Introduction of Our New Infantry Drill Regulations," *ANR* 13 (March 5, 1892), 158; Carter, "Tradition and Drill," 116. See also "Army Drill Regulations," *ANR* 15 (May 19, 1894), 325.

49. L. P. Davison, "Battle Tactics and Mounted Infantry," *JMSIUS* 20 (January–May 1897), 301; T. M. Anderson, "Comment and Criticism: Drill Regulations," 151. See also Lazelle, "Important Improvements," 371, and F. N. Maude, "The Evolution of Modern Drill Books," *JMSIUS* 14 (1893), 487.

8. *Charging against Entrenchments and Modern Rifles*

1. William E. Birkhimer, "Has the Adaptation of the Rifle-Principle to Fire-Arms Diminished the Relative Importance of Field Artillery?" *JMSIUS* 6 (1885), 204; C. A. P. Hatfield, "The Tendency of Evolution in the Army," *JMSIUS* 21 (July–November 1897), 449; William H. Carter, *From Yorktown to Santiago with the Sixth U.S. Cavalry* (Baltimore, 1900), 282.

2. Graham A. Cosmas, *An Army for Empire: The United States Army in the Spanish-American War* (Columbia, Mo., 1971), 76; AR, Report of the Adjutant General, November 1, 1898, Roll 97, 1898, vol. 1, pt. 1:253.

3. AR, Report of Maj. Gen. Wesley Merritt, August 31, 1898, Roll 97, 1898, vol. 1, pt. 2:41 (hereinafter in this chapter, all citations are to Roll 97, 1898, vol. 1, pt. 2).

4. David F. Trask, *The War with Spain in 1898* (New York and London, 1981), 371.

5. Cosmas, *Army for Empire*, 209, gives the numbers of Spanish forces in Cuba and discusses the weaknesses of their deployment. Trask, *The War with Spain*, 190, credits Shafter with 819 officers and 16,058 enlisted men.

6. Trask, *The War with Spain*, 353.

7. Graham A. Cosmas, "San Juan Hill and El Caney, 1–2 July 1898," in Charles E. Heller and William A. Stofft, eds., *America's First Battles, 1776–1965* (Lawrence, Kans., 1986), 119; Cosmas, *Army for Empire*, 160.

8. Frederick Funston, *Memories of Two Wars: Cuban and Philippine Experiences* (New York, 1911), 176, 225; Charles Johnson Post, *The Little War of Private Post* (Boston and Toronto, 1960), 131.

9. Post, *Little War,* 76; Funston, *Memories of Two Wars*, 225. See also the testimony of Maj. Gen. Wesley Merritt, *Report of the Commission Appointed by the President to Investigate the Conduct of the War Department in the War with Spain,* 8 vols. (Washington, D.C., 1899), 7:3267. Merritt noted that "what is called in military parlance . . . the 'danger space' [is] greater" with either the Mauser or the Krag than the Springfield. He expressed an overall preference for the Springfield over the other two rifles but acknowledged that "some of the line officers know more about [these shoulder arms] than I do." Ibid. On the Mauser, see also AR, Report of Maj. Gen. Joseph C. Breckinridge, July 25, 1898, 597.

10. Post, *Little War*, 76; AR, Report of Maj. Gen. Joseph C. Breckinridge, July 25, 1898, 600.

11. Robert F. Britton to My dear Dad, May 28, 1899, Robert F. Britton Papers, U.S. Army Military History Institute; Cosmas, *Army for Empire*, 158–160.

12. AR, Report of Brig. Gen. S. B. M. Young, June 29, 1898, 333; Richard Harding Davis, *The Cuban and Porto Rican Campaigns* (New York, 1898), 208. The same account appears in Richard Harding Davis, *Notes of a War Correspondent* (New York, 1911), 89.

13. Theodore Roosevelt, *The Rough Riders* (New York, 1961), 62; AR, Report of Brig. Gen. F. V. Greene, August 23, 1898, 64.

14. AR, Report of Maj. Gen. William R. Shafter, September 13, 1898, 153; James H. Wilson to J. C. Gilmore, August 9, 1898, Letters and Telegrams Sent by the First Division, First Corps, May–June 1898, Entry 69, Record Group 395, Na-

tional Archives; AR, Report of Brig. Gen. Peter C. Hains, August 15, 1898, 141; AR, Report of Brig. Gen. T. M. Anderson, August 29, 1898, 56.

15. Entry for November 2, 1899, S. B. M. Young Diary, Samuel B. M. Young Papers, U.S. Army Military History Institute; Funston, *Memories of Two Wars,* 238, 246.

16. E. S. Godfrey to C. P. Godfrey, March 27, 1898, Godfrey Family Papers, U.S. Army Military History Institute; Eli Helmick, Autobiography manuscript, Chapter 7, 74, Eli and Charles Helmick Papers, ibid.; Funston, *Memories of Two Wars,* 305, 352. On the Spanish troops as defenders, see also AR, Report of Maj. Gen. Joseph C. Breckinridge, July 25, 1898, 597 and Report of Col. Harry C. Egbert, September 2, 1898, 365.

17. Funston, *Memories of Two Wars,* 226.

18. James Parker, "The Philippine Campaign," undated manuscript in the George DuBois Coles Papers, Spanish-American War Survey, U.S. Army Military History Institute; AR, Report of Brig. Gen. Arthur MacArthur, August 22, 1898, 81. For the American losses on August 13, 1898, at Manila, see AR, Report of Brig. Gen. Arthur MacArthur, August 22, 1898, 114, and the Report of Brig. Gen. Francis V. Greene, August 23, 1898, 71. The American army conducted its operations with such success that historian Russell F. Weigley was able to write: "Tactically as well as strategically, the war did not amount to much." Russell F. Weigley, *History of the United States Army* (Bloomington, Ind., 1984), 307.

19. Davis, *Cuban and Porto Rican Campaigns,* 299; AR, Report of Maj. Gen. Nelson A. Miles, November 5, 1898, 36; Funston, *Memories of Two Wars,* 149.

20. AR, Report of Maj. Gen. Nelson A. Miles, November 5, 1898, 34–36. See also the map in Trask, *The War with Spain,* 354.

21. AR, Report of Second Lt. Celwyn E. Hampton, August 22, 1898, 85; AR, Report of Capt. S. R. Whitall, August 2, 1898, 287.

22. Davis, *Cuban and Porto Rican Campaigns,* 164, and see also 149.

23. AR, Report of Capt. Leven C. Allen, July 24, 1898, 281; Cosmas, "San Juan Hill and El Caney," 144. See also AR, Report of Col. Leonard Wood, July 6, 1898, 343, and Report of Brig. Gen. William Ludlow on the Battle of El Caney, July 4, 1898, Letters and Reports on the Battle of El Caney, Entry 565, Record Group 395, National Archives.

24. AR, Report of Brig. Gen. T. M. Anderson, August 29, 1898, 58, and Report of Capt. Stephen O'Connor, August 15, 1898, 118.

25. AR, Report of Brig. Gen. Arthur MacArthur, August 22, 1898, 80. See also AR, Report of Capt. Lea Febiger, August 15, 1898, 116.

26. This soldier's name appears as "Sgt. Andrew J. Gaughran" on the Return of the Third U.S. Artillery, December 1898, Roll 24, Returns from Regular Army Artillery Regiments, June 1821–January 1901, M-727, National Archives.

27. No sergeants named Sullivan appear in the returns of Battery K of the Third Artillery during this period. The March 1899 rolls list a Private David J. Sullivan, wounded on March 25, who might have held a higher grade half a year earlier. The battery also had a Private Martin O. Sullivan, who was discharged on January 5, 1901. Returns of the Third Artillery, March 1899 and January 1901, ibid.

28. AR, Report of Second Lt. Lloyd England, August 20, 1898, 108; Capt. James O'Hara to the Assistant Adjutant General, Second Brigade, August 25, 1898, Letters Received and Endorsements Sent by the Second Brigade, Second Division, Eighth Corps, 1898–1900, Entry 2285, Record Group 395, National Archives. This NCO may well have been Sergeant Earl Fisher, later wounded during the Philippine War. Casualty Returns of the Third U.S. Artillery, March 25–April 1, 1899, and March 25–April 13, 1899, Roll 24, Returns from Regular Army Artillery Regiments, June 1821–January 1901, M-727, National Archives. A Sergeant Frederic C. Fisher also served with the same battery until his discharge on July 27, 1899. Return of the Third U.S. Artillery, July 1899, ibid.

29. AR, Report of Captain George H. Palmer, July 18, 1898, 280; AR, Report of Brig. Gen. Arthur MacArthur, August 22, 1898, 80, 82.

30. Col. Irving Hale to the Assistant Adjutant General, Second Brigade, Second Division, Eighth Corps, August 25, 1898, Letters Received and Endorsements Sent by the Second Brigade, Second Division, Eighth Corps.

31. AR, Report of Second Lt. John H. Parker, July 23, 1898, 458; John H. Parker, *History of the Gatling Gun Detachment, Fifth Army Corps, at Santiago, with a Few Unvarnished Truths Concerning That Expedition* (Kansas City, Mo., 1898); Roosevelt, *Rough Riders*, 108, and see also 97.

32. AR, Report of Maj. Gen. William R. Shafter, September 13, 1898, 154; AR, Report of Second Lt. D. W. Ryther, July 18, 1898, 298; AR, Report of Capt. L. W. V. Kennon, July 9, 1898, 288.

33. Cosmas, *Army for Empire*, 214; AR, Report of Capt. Leven C. Allen, July 24, 1898, 281; AR, Report of First Lt. R. R. Steedman, July 31, 1898, 282; Telegram, [James H.] Wilson to [J. C.] Gilmore, August 11, 1898, Letters and Telegrams Sent by the First Division, First Corps; Funston, *Memories of Two Wars*, 182, 184. Historian Graham Cosmas believed the poor showing of the field guns at San Juan "did not shake the predominant faith in direct-fire artillery tactics." Cosmas, "San Juan Hill and El Caney," 147.

34. Trask, *The War with Spain*, 173–174, 341; AR, Report of Brig. Gen. Guy V. Henry, August 19, 1898, 246. Volunteer cavalry and a sprinkling of regular cavalry, far less than the full regiment Henry had in mind, served in Puerto Rico. Ibid., Report of Maj. Gen. John R. Brooke, August 18, 1898, 140; Report of Second Lt. William S. Valentine, August 14, 1898, 263.

35. AR, Report of Brig. Gen. Guy V. Henry, August 19, 1898, 246. Henry, whose command entered Ponce on August 1, 1898, claimed that the cavalry operation he envisioned could have been completed by August 13. AR, Report of Brig. Gen. Guy V. Henry, August 21, 1898, 248; AR, Report of Brig. Gen. Guy V. Henry, August 19, 1898, 246.

36. Roosevelt, *Rough Riders*, 135–136.

37. E. J. McClernand to S. B. M. Young, March 3, 1900, Samuel B. M. Young Papers; Edward S. Godfrey to C. P. Godfrey, November 20, 1899, Godfrey Family Papers; Funston, *Memories of Two Wars*, 240–241 (quotation, 241).

38. AR, Report of Brig. Gen. S. B. M. Young, June 29, 1898, 334; ibid., Report of Col. Harry C. Egbert, September 2, 1898, 366; Parker, *History of the Gatling Gun Detachment*, 11–12. See also Carter, *From Yorktown to Santiago*, 286.

39. Joseph Wheeler, *The Santiago Campaign, 1898* (Boston, 1898), 182. On the Santiago entrenchments, see also the entry for July 6, 1898, Diary of the Aide-de-camp of the Third Cavalry, Samuel B. M. Young Papers.

40. Thomas M. Anderson to the Adjutant General, July 9, 1898, Letterbook of the Second Division, Eighth Corps, June–August 1898, Entry 842, Record Group 395, National Archives; Telegram, Maj. Gen. James H. Wilson to the Adjutant General, August 3, 1898, Letters and Telegrams Sent by the First Division, First Corps; Telegram, [James H.] Wilson to [J. C.] Gilmore, August 11, 1898, Letters and Telegrams Sent by the First Division, First Corps. See also Telegram, Maj. Gen. James H. Wilson to Brig. Gen. J. C. Gilmore, August 4, 1898, Letters and Telegrams Sent by the First Division, First Corps.

41. Funston, *Memories of Two Wars*, 236.

42. Ibid., 184.

43. AR, Report of Brig. Gen. J. Ford Kent, July 7, 1898, 166. In front of San Juan Hill, the Americans encountered a barbed-wire fence "set mostly into living trees," which proved as difficult an obstacle as any of the prepared defenses. Ibid., Report of Col. H. A. Theaker, July 5, 1898, 279. Many official reports mention this barrier, including ibid., Report of Capt. George H. Palmer, July 18, 1898, 280; Report of First Lt. R. R. Steedman, July 31, 1898, 282; Report of Second Lt. Louis H. Bash, July 8, 1898, 431. See also Cosmas, "San Juan Hill and El Caney," 136.

For examples of the Spanish use of wire entanglements during the insurrection, see Funston, *Memories of Two Wars*, 83, 89, 91, 131, 146.

44. Wheeler, *Santiago Campaign*, 182; AR, Brig. Gen. T. M. Anderson, August 29, 1898, 56. See also AR, Report of Brig. Gen. F. V. Greene, August 23, 1898, 74.

45. Cosmas, *Army for Empire*, 124; Nelson A. Miles, *Serving the Republic: Memoirs of the Civil and Military Life of Nelson A. Miles* (Freeport, N.Y., 1971), 304; Trask, *The War with Spain*, 219; Roosevelt, *Rough Riders*, 65, 84; General Orders No. 4, March 24, 1898, Letters Received and Endorsements Sent by the Second Brigade, Second Division, Eighth Corps.

46. AR, Report of Maj. Gen. Joseph C. Breckinridge, July 25, 1898, 599; Entry for July 31, 1898, George R. Fisher Diary, photocopy in George R. Fisher Papers, Spanish-American War Survey, U.S. Army Military History Institute; Funston, *Memories of Two Wars*, 195–196. Funston had attempted the same tactics a few days earlier, but on that occasion the enemy "broke and ran" rather than "wait for the bayonet" Ibid., 185.

47. James Richard Woolard, "The Philippine Scouts: The Development of America's Colonial Army" (Ph.D. diss., Ohio State University, 1975), 2–45.

48. Ibid., 12, 66.

49. Ibid., ii, 1–2; Stephen C. Mills to My dear "Spoons" [W. W. Wotherspoon], December 1, 1908, Letterbook, 1907–1910, Stephen C. Mills Papers, U.S. Army Military History Institute.

50. Brian McAllister Linn, *The U.S. Army and Counterinsurgency in the Philippine War, 1899–1902* (Chapel Hill, N.C., and London, 1989), 81; Funston, *Memories of Two Wars*, 319; Frederick Dent Grant quoted in Woolard, "The Philippine Scouts," 76; W. P. Duvall to the Adjutant General, December 13, 1909, unsigned carbon copy in Letterbook, 1907–1910, Stephen C. Mills Papers, U.S. Army Military History Institute.

51. Trask, *The War with Spain*, 363.

52. Miles quoted in ibid., 362. On Brooke's conduct during the Puerto Rican campaign, ibid., 362–363.

53. Roosevelt, *Rough Riders*, 121; AR, Report of Maj. Gen. Wesley Merritt, August 31, 1898, 42.

54. AR, Report of Col. James S. Smith, August 18, 1898, 95; General Orders No. 4, March 24, 1898, Letters Received and Endorsements Sent by the Second Brigade, Second Division, Eighth Corps; Funston, *Memories of Two Wars*, 228, 240, 246. For another example of the advance by rushes, see ibid., 258.

55. Davis, *Cuban and Porto Rican Campaigns*, 218–220. The same account appears in Davis, *Notes of a War Correspondent*, 97–98.

56. Davis, *Cuban and Porto Rican Campaigns*, 218.

57. AR, Report of Capt. L. W. V. Kennon, July 9, 1898, 290; Post, *Little War*, 185.

58. AR, Report of Capt. Leven C. Allen, July 24, 1898, 281; Funston, *Memories of Two Wars*, 195; Thomas W. Crouch, *A Leader of Volunteers: Frederick Funston and the Twentieth Kansas in the Philippines, 1898–1899* (Lawrence, Kans., 1984), 76.

59. Funston, *Memories of Two Wars*, 199–202 (quotations, 199–200, 201). See also Crouch, *A Leader of Volunteers*, 86.

60. Funston, *Memories of Two Wars*, 247, 255.

61. George S. Patton, Jr., *War as I Knew It* (Boston, 1947), 339.

62. General Orders No. 4, March 24, 1898, Letters Received and Endorsements Sent by the Second Brigade, Second Division, Eighth Corps; Report of Lt. Col. G. S. Carpenter of the Battle of El Caney, Letters and Reports on the Battle of El Caney; AR, Report of Capt. L. W. V. Kennon, July 9, 1898, 290.

63. AR, Report of Col. Leonard Wood, July 14, 1898, 354–355; AR, Report of Sgt. Hallett Alsop Borrowe, July 14, 1898, 354.

64. Paul H. Carlson, *"Pecos Bill": A Military Biography of William R. Shafter* (College Station, Tex., 1989), 175. On the influence of the small-unit commanders, see ibid., 176–177.

65. AR, Report of Lt. Col. R. E. Thompson, August 19, 1898, 126; ibid., Report of Maj. Gen. Thomas M. Anderson, December 24, 1898, 677.

66. Brigadier General Arthur MacArthur used the telegraph to communicate with the reserve units of his brigade but not to direct his regiments in battle. AR, Report of Col. Samuel Overshine, August 17, 1898, 115.

67. These problems surface in the ARs of several signal officers: Capt. E. A. McKenna, August 17, 1898, 128; Lt. William W. Chance, n.d., 130–131; First Lt. Philip J. Perkins, August 13, 1898, 131; First Lt. Charles E. Kilbourne, n.d., 133, 135; and Second Lt. A. J. Rudd, n.d., 135–136.

68. Cosmas, "San Juan Hill and El Caney," 133–134.

69. Ibid., 133; Frank L. Milward, "An Account of Experiences in the Spanish-American War as Recalled by Frank L. Milward, of Sixth Cavalry, Troop 'C,'" undated typescript in the Frank L. Milward Papers, Spanish-American War Survey, U.S. Army Military History Institute; Post, *Little War*, 169–170, 171.

70. Cosmas, "San Juan Hill and El Caney," 125.

71. Trask, *The War with Spain*, 236.

72. Cosmas, "San Juan Hill and El Caney," 127, 143.

73. Ibid., 112, 119–120 (quotations, 119 and 120).

74. Post, *Little War*, 185; AR, Report of Capt. L. W. V. Kennon, July 9, 1898, 290. Kennon reported that his company made the charge up San Juan Hill from "a sheltered position in the dead space under the hill." Ibid., 288.

75. AR, Report of Captain Charles Byrne, July 27, 1898, 300. On the subject of the military crest, see also Carter, *From Yorktown to Santiago*, 290.

76. Cosmas, "San Juan Hill and El Caney," 144; AR, Report of Maj. Gen. Joseph Wheeler, July 7, 1898, 174. A table in Wheeler's report, 173, shows twenty-nine of the division's officers were wounded, but the accounting on p. 174 lists only twenty-six names.

77. AR, Report of Maj. Gen. Joseph Wheeler, 173; AR, Report of Brig. Gen. J. Ford Kent, July 7, 1898, 166; Cosmas, "San Juan Hill and El Caney," 136; AR, Report of Col. Harry C. Egbert, September 2, 1898, 366; entry for July 1, 1898, John E. Woodward Diary, John E. Woodward Papers, Spanish-American War Survey, U.S. Army Military History Institute.

78. Cosmas, "San Juan Hill and El Caney," 144–145 (Roosevelt quotation, 145). See also Trask, *The War with Spain*, 248.

79. AR, Report of Lt. Col. William H. Bisbee, August 11, 1898, 382.

80. Report of Brig. Gen. William Ludlow on the Battle of El Caney, July 4, 1898, Letters and Reports on the Battle of El Caney; Frank L. Milward, "An Account of Experiences," Frank L. Milward Papers; Sixteenth Infantry officer quoted in Cosmas, "San Juan Hill and El Caney," 145.

81. Parker, *History of the Gatling Gun Detachment*, 92.

Select Bibliography

Primary Sources

Weapons

Birkhimer, William E. *Historical Sketch of the Organization, Administration, Matériel, and Tactics of the Artillery, United States Army*. Washington, D.C.: James J. Chapman, 1884.

Board of Officers. *Reports of Experiments with Rice's Trowel Bayonet, Made by Officers of the Army, Pursuant to Instructions from the War Department*. Springfield, Mass.: National Armory, 1874.

Harris, Moses. "Smokeless Powder in Its Relation to Cavalry Efficiency." *Journal of the United States Cavalry Association* 6 (1893):36–47.

Norton, Charles B. *American Inventions and Improvements in Breech-Loading Small Arms, Heavy Ordnance, Machine Guns, Magazine Arms, Fixed Ammunition, Pistols, Projectiles, Explosives, and Other Munitions of War, including a Chapter on Sporting Arms*. Springfield, Mass.: Chapin & Gould, 1880.

Walke, Willoughby. "A New Powder." *Journal of the United States Artillery* 2 (1893):374–379.

Papers of Boards of Officers on Tactics

Clitz Board Papers, AGO File 312 A 680, Roll 680, M-619, National Archives.
Correspondence on the Proposal of Captain R. P. Hughes, AGO File 310 AGO
    1884, Roll 254, M-689, National Archives.
Emory Board Papers, AGO File 312 A 1869, Roll 680, M-619, National Archives.
Grant Board Papers, AGO File 312 A 680, Roll 680, M-619, National Archives.
Proceedings and Report of the Barry Board, AGO File 312 A 1869, Rolls 681 and
    682, M-619, National Archives.
Recommendations of Lieutenant Colonel James W. Forsyth, AGO Files 1558
    AGO 1882, 5302 AGO 1882, and 2331 AGO 1882, Roll 94, M-689, National
    Archives.
Records of the Board on the Revision of Tactics, AGO File 526 AGO 1880, Rolls
    587–608, M-689, National Archives.
Schofield Board Papers, AGO File 312 A 1869, Rolls 682–685, M-619, National
    Archives.

Drill Books, Manuals, and Works on Tactical Theory

Blunt, Stanhope E. *Firing Regulations for Small Arms for the United States Army.*
    New York: Charles Scribner's Sons, 1889, 1894, 1897.
————. *Instructions in Rifle and Carbine Firing for the United States Army.* New
    York: Charles Scribner's Sons, 1885, 1886.
Cooke, Philip St. George. *Cavalry Tactics; or, Regulations for the Instruction, Forma-*
    tions, and Movements of the Army and Volunteers of the United States. 2 parts.
    Washington, D.C.: Government Printing Office, 1861, 1862. Philadelphia: J. B.
    Lippincott, 1864. New York: D. Van Nostrand, 1872.
Dyer, A. B. *Handbook for Light Artillery.* New York: John Wiley & Sons, 1896,
    1900.
Farrow, Edward S. *Mountain Scouting: A Hand-Book for Officers and Soldiers on the*
    Frontiers. New York: published by the author, 1881.
French, John T., Jr., ed. *Interpretations of Infantry Drill Regulations, United States*
    Army. New York: Army and Navy Journal, 1893.
————. *Revision of Interpretations of Infantry Drill Regulations.* New York and
    Washington, D.C.: Army and Navy Journal and James J. Chapman, 1893.
Gibson, George. *Battalion Drill and Battalion Skirmish Drill Compiled from Upton's*
    Infantry Tactics and War Department Decisions. Fort Keough, Montana Terri-
    tory: published by the author, 1887.

Hamley, Bruce Edward. *The Operations of War*. Edinburgh and London: William Blackwood and Sons, 1866.

Home, Robert. *Précis of Modern Tactics*. London: Her Majesty's Stationery Office, 1882.

———. *A Précis of Modern Tactics*. Edited by Sisson C. Pratt. London: Her Majesty's Stationery Office, 1896.

Kelton, J. C. *Information for Riflemen on the Range and the Battle-field*. San Francisco: published by the author, 1884.

Laidley, T. T. S. *A Course of Instruction in Rifle Firing*. Philadelphia: J. B. Lippincott & Company, 1879, 1880.

Lippitt, Francis J. *A Treatise On Intrenchments*. New York: D. Van Nostrand, 1866.

———. *A Treatise on the Tactical Use of the Three Arms: Infantry, Artillery, and Cavalry*. New York: D. Van Nostrand, 1865.

Mayne, C. B. *Infantry Fire Tactics*. Chatham, England: Gale and Polden, Brompton Works, 1885.

Morris, William H. *Field Tactics for Infantry*. New York: D. Van Nostrand, 1864.

———. *Infantry Tactics*. New York: D. Van Nostrand, 1865.

———. *Tactics for Infantry Armed with Breech-Loading or Magazine Rifles*. New York: American Bank Note Company, 1882.

Patterson, John H. *Infantry Tactics: A System of Attack and Defence With Re-enforced Skirmish Lines, for the Instruction and Guidance of Officers and Non-Commissioned Officers; School of the Company and Battalion*. Fort Leavenworth, Kans.: U.S. Infantry and Cavalry School, 1884.

Reed, Hugh T. *Abridgement of the Drill Regulations for Infantry*. Chicago: published by the author, 1891.

———. *Elements of Military Science and Tactics*. Chicago: published by the author, 1889.

Tidball, John C. *Manual of Heavy Artillery Service Prepared for the Use of the Army and Militia of the United States*. Washington, D.C.: James J. Chapman, 1880, 1891, 1898.

Upton, Emory. *Infantry Tactics, Double and Single Rank*. New York: D. Appleton, 1874, 1875, 1876, 1877, 1884, 1885, 1886, 1887, 1889, 1890.

———. *A New System of Infantry Tactics, Double and Single Rank, Adapted to American Topography and Improved Fire-Arms*. New York: D. Appleton, 1867, 1868, 1869.

United States War Department. *Artillery Tactics, United States Army, Assimilated to the Tactics of Infantry and Cavalry*. New York: D. Appleton, 1874, 1876, 1877, 1878, 1882, 1883, 1885, 1886.

―――. *Cavalry Drill Regulations, United States Army.* Washington, D.C.: Government Printing Office, 1891.

―――. *Cavalry Tactics, United States Army, Assimilated to the Tactics of Infantry and Artillery.* New York: D. Appleton, 1874, 1878, 1879, 1883.

―――. *Drill Regulations for Cavalry, United States Army.* Washington, D.C.: Government Printing Office, 1896.

―――. *Drill Regulations, Light Artillery, United States Army.* Washington, D.C.: Government Printing Office, 1896.

―――. *Infantry Drill Regulations: The Manual of Arms, Adapted to the Magazine Rifle, Caliber .30.* Washington, D.C.: Government Printing Office, 1897.

―――. *Infantry Drill Regulations, United States Army.* Washington, D.C.: Government Printing Office, 1891. New York: D. Appleton, 1891, 1892, 1893, 1895, 1898. New York: D. Appleton, Army and Navy Journal, 1893, 1898.

―――. *Light Artillery Drill Regulations, United States Army.* Washington, D.C.: Government Printing Office, 1891.

―――. *Regulations for the Army of the United States, 1889.* Washington, D.C.: Government Printing Office, 1889.

―――. *Regulations of the Army of the United States and General Orders in Force on the 17th of February, 1881.* Washington, D.C.: Government Printing Office, 1881.

Wheeler, Junius Brutus. *The Elements of Field Fortifications for the Use of the Cadets of the United States Military Academy at West Point, N.Y.* New York: D. Van Nostrand, 1882.

Wingate, George W. *Manual for Rifle Practice.* New York: W. C. and F. P. Church, 1875.

Official Reports

Report of the Commission Appointed by the President to Investigate the Conduct of the War Department in the War with Spain. 8 vols. Washington, D.C.: Government Printing Office, 1899.

United States War Department. *Annual Reports of the War Department, 1822–1907.* M-997, National Archives.

―――. *The War of the Rebellion: A Compilation of the Official Records of the Union and Confederate Armies.* 128 vols. Washington, D.C.: Government Printing Office, 1880–1901.

Army Journals

Army and Navy Journal, 1865–1897.
Army and Navy Register, 1882–1897.
Journal of the Military Service Institution of the United States, 1880–1897.
Journal of the United States Artillery, 1892–1897.
Journal of the United States Cavalry Association, 1888–1897.

Papers of General Officers

Tasker H. Bliss Papers, United States Army Military History Institute.
Godfrey Family Papers, United States Army Military History Institute.
Guy V. Henry, Sr., Papers, United States Army Military History Institute.
Ranald S. Mackenzie Papers, United States Army Military History Institute.
Nelson A. Miles Family Papers, Library of Congress.
John M. Schofield Papers, Library of Congress.
Philip H. Sheridan Papers, Library of Congress.
William T. Sherman Papers, Library of Congress.
Samuel B. M. Young Papers, United States Army Military History Institute.

Memoirs and Recollections

Carter, William H. *From Yorktown to Santiago with the Sixth U.S. Cavalry*. Baltimore: Lord Baltimore Press, 1900.
Cox, Jacob D. *Military Reminiscences of the Civil War*. 2 vols. New York: Charles Scribner's Sons, 1900.
Hood, John B. *Advance and Retreat*. Bloomington, Ind.: Indiana University Press, 1959.
Hutcheson, Grote. Diary. Grote Hutcheson Papers, United States Army Military History Institute.
Johnson, Robert Underwood, and Clarence Clough Buell, eds. *Battles and Leaders of the Civil War*. 4 vols. 1884–1888. Reprint. New York and London: Thomas Yoseloff, 1956.
Marcy, Randolph B. *Thirty Years of Army Life on the Border*. 1866. Reprint. Philadelphia and New York: J. B. Lippincott, 1963.

Miles, Nelson A. *Personal Recollections and Observations of General Nelson A. Miles.* Chicago and New York: Werner , 1896.

————. *Serving the Republic: Memoirs of the Civil and Military Life of Nelson A. Miles.* 1911. Reprint. Freeport, N.Y.: Books for Libraries Press, 1971.

Mills, Anson. *My Story.* Edited by C. H. Claudy. Washington, D.C.: published by the author, 1918.

Schmitt, Martin F., ed. *General George Crook: His Autobiography.* London and Norman, Okla.: University of Oklahoma Press, 1986.

Sheridan, Philip H. *Personal Memoirs of P. H. Sheridan.* 2 vols. New York: Charles L. Webster, 1888.

Manuscripts on the Indian Wars

Luther P. Bradley Papers, United States Army Military History Institute.

Chiricahua Apache Papers, AGO File 1066 AGO 1883, Rolls 173–188, M-689, National Archives.

Crook-Kennon Papers, United States Army Military History Institute.

Fetterman Papers, AGO File 102 M 1867, Roll 560, M-619, National Archives.

Gray-Woodruff Papers, United States Army Military History Institute.

Little Big Horn Papers, AGO Files 3765-3776 AGO 1876, Roll 273, M-666, National Archives.

Military Division of the Missouri Papers, AGO Files 102–196 R 1870, Roll 812, M-619, National Archives.

Military Division of the Missouri Papers, United States Army Military History Institute.

Military Order of the Loyal Legion of the United States Collection, United States Army Military History Institute.

Stephen C. Mills Papers, United States Army Military History Institute.

Order of the Indian Wars Papers, United States Army Military History Institute.

Reports of Generals Hancock and Custer against the Sioux and Cheyenne, AGO File 590 M 1867, Roll 563, M-619, National Archives.

Rosebud Papers, AGO File 3570 AGO 1876, Roll 271, M-666, National Archives.

Publications on the Indian Wars

Bourke, John G. *An Apache Campaign in the Sierra Madre.* 1886. Reprint. New York: Charles Scribner's Sons, 1958.

————. *Mackenzie's Last Fight with the Cheyennes: A Winter Campaign in Wyoming and Montana.* 1890. Reprint. Bellevue, Nebr.: Old Army Press, 1970.

————. *On the Border with Crook.* 1891. Reprint. Lincoln: University of Nebraska Press, 1971.

Bradley, James H. *The March of the Montana Column: A Prelude to the Custer Disaster.* Edited by Edgar I. Stewart. Norman: University of Oklahoma Press, 1961.

Brininstool, E. A. *Troopers with Custer: Historic Incidents of the Battle of the Little Big Horn.* London and Lincoln, Nebr.: University of Nebraska Press, 1989.

Carroll, John M., ed. *The Papers of the Order of the Indian Wars.* Fort Collins, Colo.: Old Army Press, 1975.

Clark, Keith, and Donna Clark, eds. "William McKay's Journal, 1866–67: Indian Scouts, Part I." *Oregon Historical Quarterly* 79:2 (Summer 1978):121–171.

————. "William McKay's Journal, 1866–67: Indian Scouts, Part II." *Oregon Historical Quarterly* 79:3 (Fall 1978):268–333.

Finerty, John F. *War-Path and Bivouac: The Big Horn and Yellowstone Expedition.* Edited by Milo Milton Quaife. London and Lincoln: University of Nebraska Press, 1966.

Gibbon, John. "Arms to Fight Indians." *United Service* 1 (April 1879):237–244.

Graham, W. A. *The Custer Myth: A Source Book of Custeriana.* New York: Bonanza Books, 1958.

Hammer, Kenneth, ed. *Custer in '76: Walter Camp's Notes on the Custer Fight.* Provo, Utah: Brigham Young University Press, 1976.

Keim, De B. Randolph. *Sheridan's Troopers on the Borders: A Winter Campaign on the Plains.* 1885. Reprint. London and Lincoln: University of Nebraska Press, 1985.

King, Charles. *Campaigning with Crook.* Norman: University of Oklahoma Press, 1964.

McClernand, Edward J. *With the Indian and the Buffalo in Montana, 1870–1878.* Glendale, Calif.: Arthur H. Clark Company, 1969.

Utley, Robert M., ed. *Life in Custer's Cavalry: Diaries and Letters of Albert and Jennie Barnitz, 1867–1868.* London and New Haven, Conn.: Yale University Press, 1977.

Spanish-American and Philippine Wars

Robert F. Britton Papers, United States Army Military History Institute.

George DuBois Coles Papers, Spanish-American War Survey, United States Army Military History Institute.

Davis, Richard Harding. *The Cuban and Porto Rican Campaigns*. New York: Charles Scribner's Sons, 1898.

————. *Notes of a War Correspondent*. New York: Charles Scribner's Sons, 1911.

George R. Fisher Papers, Spanish-American War Survey, United States Army Military History Institute.

Funston, Frederick. *Memories of Two Wars: Cuban and Philippine Experiences*. New York: Charles Scribner's Sons, 1911.

Hard, Curtis V. *Banners in the Air: The Eighth Ohio Volunteers and the Spanish-American War*. Edited by Robert H. Ferrell. London and Kent, Ohio: Kent State University Press, 1988.

Eli and Charles Helmick Papers, United States Army Military History Institute.

Letterbook of the Second Division, Eighth Corps, June–August 1898, Entry 842, Record Group 395, National Archives.

Letters Received and Endorsements Sent by the Second Brigade, Second Division, Eighth Corps, 1898–1900, Entry 2285, Record Group 395, National Archives.

Letters and Reports on the Battle of El Caney, Entry 565, Record Group 395, National Archives.

Letters and Telegrams Sent by the First Division, First Corps, May–June 1898, Entry 69, Record Group 395, National Archives.

Frank L. Milward Papers, Spanish-American War Survey, United States Army Military History Institute.

Parker, John H. *History of the Gatling Gun Detachment, Fifth Army Corps, at Santiago, with a Few Unvarnished Truths Concerning that Expedition*. Kansas City, Mo.: Hudson-Kimberly, 1898.

Post, Charles Johnson. *The Little War of Private Post*. Boston and Toronto: Little, Brown, 1960.

Roosevelt, Theodore. *The Rough Riders*. New York: New American Library, Signet Classic, 1961.

Wheeler, Joseph. *The Santiago Campaign, 1898*. Boston: Lamson, Wolffe, 1898.

John E. Woodward Papers, Spanish-American War Survey, United States Army Military History Institute.

Secondary Sources

Weapons

Armstrong, David A. *Bullets and Bureaucrats: The Machine Gun and the United States Army, 1861–1916.* London and Westport, Conn.: Greenwood Press, 1982.

Brinckerhoff, Sidney B., and Pierce Chamberlain. "The Army's Search for a Repeating Rifle, 1873–1903." *Military Affairs* 32 (April 1968):20–30.

Bruce, Robert V. *Lincoln and the Tools of War.* Indianapolis: Bobbs-Merrill, 1956.

Schrier, Konrad F. "U.S. Army Field Artillery Weapons, 1866–1918." *Military Collector and Historian* 20 (Summer, 1968):40–45.

Wolf, Richard I. "Arms and Innovation: The United States Army and the Repeating Rifle, 1865–1900." Ph.D. diss., Boston University, 1981.

Tactics

Gilmore, Russell. "'The New Courage': Rifles and Soldier Individualism, 1876–1918." *Military Affairs* 40:3 (October 1976):97–102.

Griffith, Paddy. *Battle Tactics of the Civil War.* London and New Haven, Conn.: Yale University Press, 1989.

Hagerman, Edward. *The American Civil War and the Origins of Modern Warfare: Ideas, Organization, and Field Command.* Indianapolis and Bloomington, Ind.: Indiana University Press, 1988.

Jamieson, Perry D. "The Development of Civil War Tactics." Ph.D. diss., Wayne State University, 1979.

McWhiney, Grady, and Perry D. Jamieson. *Attack and Die: Civil War Military Tactics and the Southern Heritage.* Tuscaloosa, Ala.: University of Alabama Press, 1982.

Nesmith, Vardell Edwards, Jr. "The Quiet Paradigm Change: Evolution of the Field Artillery Doctrine of the United States Army, 1861–1905." Ph.D. diss., Duke University, 1977.

Indian Wars

Du Bois, Charles G. *The Custer Mystery.* El Segundo, Calif.: Upton and Sons, 1986.

Dunlay, Thomas W. *Wolves for the Blue Soldiers: Indian Scouts and Auxiliaries with* the United States Army, 1860–90. London and Lincoln: University of Nebraska Press, 1987.

Essin, Emmett M., III. "Notes and Documents: Mules, Packs, and Packtrains." *Southwestern Historical Quarterly* 74:1 (July 1970):52–82.

Graham, W. A. *The Story of the Little Big Horn: Custer's Last Fight.* London and Lincoln: University of Nebraska Press, 1988.

Gray, John S. *The Centennial Campaign: The Sioux War of 1876.* London and Norman: Oklahoma University Press, 1988.

———. *Custer's Last Campaign: Mitch Boyer and the Little Bighorn Reconsidered.* London and Lincoln: University of Nebraska Press, 1991.

Greene, Jerome A. *Slim Buttes, 1876: An Episode of the Great Sioux War.* Norman: University of Oklahoma Press, 1982.

———. *Yellowstone Command: Colonel Nelson A. Miles and the Great Sioux War, 1876–1877.* London and Lincoln: University of Nebraska Press, 1991.

Grinnell, George Bird. *The Fighting Cheyennes.* 1915. Reprint. Norman: University of Oklahoma Press, 1971.

———. *Two Great Scouts and Their Pawnee Battalion.* Cleveland: Arthur H. Clark, 1928.

Hammer, Kenneth M. "The Glory March: A Concise Account of the Little Bighorn Campaign of 1876." *English Westerners' Brand Book* 8:4 (July 1966):1–6.

Hebard, Grace Raymond, and E. A. Brininstool. *The Bozeman Trail: Historical Accounts of the Blazing of the Overland Routes into the Northwest, and the Fights with Red Cloud's Warriors.* 2 vols. Cleveland: Arthur H. Clark, 1922.

Hedren, Paul L., ed. *The Great Sioux War, 1876–77.* Helena: Montana Historical Society Press, 1991.

Hutchins, James S. "Mounted Riflemen: The Real Role of Cavalry in the Indian Wars." *El Palacio* 69 (Summer 1962):85–91.

Mangum, Neil C. *Battle of the Rosebud: Prelude to the Little Bighorn.* El Segundo, Calif.: Upton and Sons, 1987.

Mason, Joyce Evelyn. "The Use of Indian Scouts in the Apache Wars, 1870–1886." Ph.D. diss., Indiana University, 1970.

Nye, Wilbur Sturdevant. *Plains Indian Raiders: The Final Phases of Warfare from the Arkansas to the Red River.* Norman: University of Oklahoma Press, 1974.

Porter, Joseph C. *Paper Medicine Man: John Gregory Bourke and His American West.* London and Norman: University of Oklahoma Press, 1986.

Rickey, Don, Jr. *Forty Miles a Day on Beans and Hay: The Enlisted Soldier Fighting the Indian Wars.* Norman: University of Oklahoma Press, 1977.

———. *War in the West: The Indian Campaigns*. Fort Collins, Colo.: Old Army Press, 1956.

Shields, G. O. *The Battle of the Big Hole*. Chicago and New York: Rand, McNally, 1899.

Smith, Sherry L. *The View from Officers' Row: Army Perceptions of Western Indians*. Tucson: University of Arizona Press, 1990.

Stewart, Edgar I. *Custer's Luck*. Norman: University of Oklahoma Press, 1955.

Stout, Joseph A., Jr. *Apache Lightning: The Last Great Battles of the Ojo Calientes*. New York: Oxford University Press, 1974.

Thrapp, Dan L. *General Crook and the Sierra Madre Adventure*. Norman: University of Oklahoma Press, 1972.

Utley, Robert M. *Custer and the Great Controversy: The Origin and Development of a Legend*. Pasadena: Westernlore Press, 1980.

———. *Frontier Regulars: The United States Army and the Indian, 1866–1891*. New York: Macmillan, 1973.

———. *The Last Days of the Sioux Nation*. London and New Haven, Conn.: Yale University Press, 1963.

Vaughn, J. W. *The Reynolds Campaign on Powder River*. Norman: University of Oklahoma Press, 1961.

———. *With Crook at the Rosebud*. London and Lincoln: University of Nebraska Press, 1988.

Werner, Fred H. *The Dull Knife Battle*. Greeley, Colo.: Werner, 1981.

Whitman, S. E. *The Troopers: An Informal History of the Plains Cavalry, 1865–1890*. New York: Hastings House Publishers, 1962.

Wooster, Robert. *The Military and United States Indian Policy, 1865–1903*. London and New Haven, Conn.: Yale University Press, 1988.

Spanish-American and Philippine Wars

Cosmas, Graham A. *An Army for Empire: The United States Army in the Spanish-American War*. Columbia: University of Missouri Press, 1971.

———. "San Juan Hill and El Caney, 1–2 July 1898." In *America's First Battles, 1776–1965*, edited by Charles E. Heller and William A. Stofft. Lawrence: University Press of Kansas, 1986.

Early, Gerald H. "The United States Army in the Philippine Insurrection, 1899–1902." Master's thesis, U.S. Army Command and General Staff College, 1975.

Gates, John Morgan. *Schoolbooks and Krags: The United States Army in the Philippines, 1898–1902*. London and Westport, Conn.: Greenwood Press, 1973.

Linn, Brian McAllister. *The U.S. Army and Counterinsurgency in the Philippine War, 1899–1902*. London and Chapel Hill, N.C.: University of North Carolina Press, 1989.

Millis, Walter. *The Martial Spirit: A Study of Our War with Spain*. Boston and New York: Houghton Mifflin, 1931.

O'Toole, G. J. A. *The Spanish War: An American Epic—1898*. New York: W. W. Norton, 1984.

Roth, Russell. *Muddy Glory: America's "Indian Wars" in the Philippines, 1899–1935*. West Hanover, Mass.: Christopher, 1981.

Trask, David F. *The War with Spain in 1898*. New York: Macmillan, 1981.

Welch, Richard E., Jr. *Response to Imperialism: The United States and the Philippine-American War, 1899–1902*. Chapel Hill: University of North Carolina Press, 1979.

Wolff, Leon. *Little Brown Brother: How the United States Purchased and Pacified the Philippine Islands at the Century's Turn*. Garden City, N.Y.: Doubleday, 1961.

Woolard, James Richard. "The Philippine Scouts: The Development of America's Colonial Army." Ph.D. diss., Ohio State University, 1975.

Biographies

Ambrose, Stephen E. *Upton and the Army*. Baton Rouge: Louisiana State University Press, 1964.

Bowie, Chester Winston. "Redfield Proctor: A Biography." Ph.D. diss., University of Wisconsin–Madison, 1980.

Carlson, Paul H. *"Pecos Bill": A Military Biography of William R. Shafter*. College Station: Texas A & M University Press, 1989.

Crouch, Thomas W. *A Leader of Volunteers: Frederick Funston and the Twentieth Kansas in the Philippines, 1898–1899*. Lawrence, Kans.: Coronado Press, 1984.

Hutton, Paul Andrew. *Phil Sheridan and His Army*. London and Lincoln: University of Nebraska Press, 1985.

Michie, Peter S. *The Life and Letters of Emory Upton*. New York: D. Appleton, 1885.

Morris, Roy, Jr. *Sheridan: The Life and Wars of General Phil Sheridan*. New York: Crown, 1992.

Russell, Don. *Campaigning with King: Charles King, Chronicler of the Old Army.* Edited by Paul L. Hedren. London and Lincoln: University of Nebraska Press, 1991.

Utley, Robert M. *Cavalier in Buckskin: George Armstrong Custer and the Western Military Frontier.* London and Norman: University of Oklahoma Press, 1988.

Reference Works

Cullum, George W. *Biographical Register of the Officers and Graduates of the U.S. Military Academy at West Point, New York.* 9 vols. Cambridge, Mass.: Riverside Press, 1901.

Dawson, Joseph G., III. *The Late 19th Century U.S. Army, 1865–1898: A Research Guide.* New York, London, and Westport, Conn.: Greenwood Press, 1990.

Heitman, Francis B. *Historical Register and Dictionary of the United States Army, from Its Organization, September 29, 1789, to March 2, 1903.* 2 vols. Washington, D.C.: Government Printing Office, 1903.

Spiller, Roger J., Joseph G. Dawson III, and T. Harry Williams, eds. *Dictionary of American Military Biography.* 3 vols. London and Westport, Conn.: Greenwood Press, 1984.

Unsworth, Michael E. "Army Service Journals, 1879–1917: Vehicles of Professionalism." Paper presented at the fifty-sixth annual meeting of the American Military Institute, Lexington, Va., April 15, 1989.

Other Publications

Bigelow, Donald Nevius. *William Conant Church and the Army Navy Journal.* New York: Columbia University Press, 1952.

Coffman, Edward M. *The Old Army: A Portrait of the American Army in Peacetime, 1784–1898.* New York and Oxford: Oxford University Press, 1986.

Fuller, J. F. C. *A Military History of the Western World.* 3 vols. New York: Funk and Wagnalls, 1954–1956.

Jones, Archer. *The Art of War in the Western World.* Chicago and Urbana: University of Illinois Press, 1987.

Leach, Douglas Edward. *Arms for Empire: A Military History of the British Colonies in North America, 1607–1763.* New York and London: Macmillan and Collier-Macmillan, 1973.

————. *Flintlock and Tomahawk: New England in King Philip's War*. New York: W. W. Norton, Norton Library, 1966.

Leonard, Thomas C. *Above the Battle: War-Making in America from Appomattox to Versailles*. New York: Oxford University Press, 1978.

Luvaas, Jay. *The Military Legacy of the Civil War: The European Inheritance*. Chicago: University of Chicago Press, 1959.

Mahon, John K. *History of the Militia and the National Guard*. New York and London: Macmillan, 1983.

————. *History of the Second Seminole War, 1835–1842*. Rev. ed. Gainesville, Fla.: University Presses of Florida, University of Florida Press, 1985.

Millet, Allan R., and Peter Maslowski. *For the Common Defense: A Military History of the United States of America*. New York and London: Free Press, 1984.

Nenninger, Timothy K. *The Leavenworth Schools and the Old Army: Education, Professionalism, and the Officer Corps of the United States Army, 1881–1918*. London and Westport, Conn.: Greenwood Press, 1978.

Reardon, Carol. *Soldiers and Scholars: The U.S. Army and the Uses of Military History, 1865–1920*. Lawrence: University Press of Kansas, 1990.

Sefton, James E. *The United States Army and Reconstruction, 1865–1877*. Baton Rouge: Louisiana State University Press, 1967.

Wagner, Arthur L. *The Campaign of Königgrätz: A Study of the Austro-Prussian Conflict in Light of the American Civil War*. Kansas City, Mo.: Hudson-Kimberly, 1899.

Weigley, Russell F. *The American Way of War: A History of United States Military Strategy and Policy*. New York and London: Macmillan and Collier-Macmillan, 1973.

————. *History of the United States Army*. New York and London: Macmillan and Collier-Macmillan, 1967.

Index

About the Author

PERRY D. JAMIESON is Historian at the Center for Air Force History, Bolling Air Force Base, District of Columbia, and is the coauthor, with Grady McWhiney, of *Attack and Die: Civil War Military Tactics and the Southern Heritage.*